THE
KINFOLK
GARDEN

THE KINFOLK

KINFOLK

GARDEN

HOW *to* LIVE WITH NATURE

JOHN BURNS

ARTISAN

EDITOR IN CHIEF
John Burns

CREATIVE DIRECTOR
Staffan Sundström

EDITOR
Harriet Fitch Little

PRODUCTION MANAGER
Susanne Buch Petersen

EDITORIAL ASSISTANT
Gabriele Dellisanti

COVER PHOTOGRAPHY
Sarah Blais

BOOK DESIGN
Staffan Sundström &
Julie Freund-Poulsen

ILLUSTRATIONS
Courtesy of Rijksmuseum

Selected Contributors

Sarah Blais

Sarah is a photographer based in
Paris. She received the *British Journal
of Photography*'s Female in Focus
award in 2019.

Rodrigo Carmuega

Rodrigo is a fashion and portrait
photographer. Born in Buenos
Aires, Rodrigo has been based in
London since 2012.

Zoltán Tombor

Zoltán is a photographer in
New York and London. His work is
part of the collection at the
Hungarian Museum of Photography.

Lauren Boudreau

Lauren is a Californian florist and
stylist based in Copenhagen. She
gathers her materials and design
ideas from local nature.

Alexander Wolfe

Alexander is a photographer based
in Cape Town. His work has been
featured in the *New York Times*, *Vogue
Arabia* and *Monocle*.

Darryl Cheng

Darryl is the founder of *House Plant
Journal* and author of
The New Plant Parent, both offering
advice for indoor gardeners.

Melissa Mabbitt

Melissa is a British freelance gardening writer. She holds a qualification
in horticulture from the UK's Royal
Horticultural Society.

Amy Merrick

Amy is a traveling writer,
floral stylist and the author of
*On Flowers: Lessons from an
Accidental Florist*.

FOR FULL LIST OF CREDITS SEE PAGE 351

"Gardens are artworks that escape the control of their maker."

ABDERRAZAK BENCHAÂBANE

CONTENTS

PART THREE

COMMUNITY

INTRODUCTION

Nurture begets nature, and vice versa: It may only inspire and nourish us if we care for it in turn.

Small acts of care carry a larger significance, given the collective effort required to improve the health of the planet. Buoying this effort is our understanding that a strong connection with nature improves individual well-being. As the American essayist, and evergreen nature lover, Henry David Thoreau wrote: "There are moments when all anxiety and stated toil are becalmed in the infinite leisure and repose of nature." In Japan, *shinrin-yoku*, or "forest bathing" (quite simply lying down to meditate among the trees), is incorporated into the country's healthcare program as an antidote to the ills of modern life. The practice may sound unconventional, but it's just a different path to the same comfort we feel when walking the long route to work through the park, or on observing a plant recover after an overdue watering.

In *The Kinfolk Garden*, stories about nature as nourishment recur amid the people who graciously invited us into their gardens, studios and communities all over the world. "Caring for plants is the best way to learn how to care for yourself," says Monai Nailah McCullough, a horticulturist who leads workshops on plant care in Amsterdam. Almost 6,000 miles (10,000 km) away in Cape Town, Gundula Deutschlander—the head gardener at Babylonstoren farm—settles on an unusual, but telling, adjective to describe her relationship with nature: "Gardening has been incredibly *generous*," she says. "It makes me feel so alive."

Perhaps your garden looks nothing like those featured. Quite possibly, you currently don't have access to an outdoor space at all. Don't worry. One of the most rewarding things about plants and flowers is that their beauty will remain the same in whichever home they find themselves: And there is as much awe to be found in a slowly unfurling fern seedling on a windowsill as there is in an extravagantly planted garden. Artist Sourabh Gupta, who spends weeks studying particular species while making botanically accurate replicas out of paper, says, "There is something magnificent about a whole plant. It almost feels like poetry."

The stories in this book offer ideas for incorporating more nature into your life. We meet creative people, like Gupta, whose visionary projects are inspired by the beauty of flowers, we explore the inspiring ways in which people around the world are caring for plant life, and we join communities that are blossoming around nature. There are also practical tips throughout the book, offering advice on everything from creating beautiful flower arrangements to growing your own produce and simply keeping your peace lily alive for a little longer than usual. Many of the people we meet along the way champion the idea of following one's natural instincts. As Copenhagen's most creative florist, Julius Værnes Iversen, reminds us: "You don't need an education in this line of work. I believe that people should just jump into it and give it a go."

PART ONE

Care

Nature always finds a way to thrive. Meet those who are making it their mission to tame its wild spirit with tender loving care.

FEM GÜÇLÜTÜRK

In 2014, Fem Güçlütürk left the PR company she'd founded to retrain as a botanist in her retirement. In a quiet corner of Turkey, she discovered that caring for a busy greenhouse and garden was a conduit to caring for herself.

"This is the part where people say, 'And they lived happily ever after,'" says Fem Güçlütürk, speaking from the home she shares with her husband, Sezer Savaşli, in southwest Turkey. Since trading city life for a plot of land remote enough to lack reliable phone service, Güçlütürk has found that her days now follow the circadian rhythms of her plants. The PR executive-turned-botanist rises at six a.m. ("Even the dog doesn't wake up then," she says) and studies permaculture and edible gardening. After breakfast, she heads into the garden and remains there, weeding and pruning, until sundown. "I live in a vegetative state," she jokes.

Born in Ankara and raised in Istanbul, Güçlütürk has always cultivated an unconventional path. She worked first in bars—a rarity for women in the 1980s—before cofounding a public relations firm. Despite achieving success, she found herself increasingly disillusioned with the relentless consumerism that accompanied urban life. "Growing up in cities, we've lost our connection with nature and found ourselves in a huge global story of consumption," she says. When she stopped wanting to attend her own events, she knew it was time to quit.

In 2014, Güçlütürk announced her next iteration as a botanist. Always interested in plants, she had started to attend gardening school and run a home shop in Istanbul; three years later, she'd relocated to Muğla, a province that boasts a rich and diverse habitat. *Celtis australis*, a tree native to the region, shares its Turkish name, *Çıtlık*, with the nearest village. "The tree has its place in mythology," says Güçlütürk. "They say that if you eat the tree's berries once, you can't leave the place, which is true! I really don't want to go anywhere else."

Before she relocated to Muğla, Güçlütürk was a voracious explorer, touring the world on her motorbike. But since she "designed her own heaven," she's loath to leave it. Instead, from the glass-fronted house that she and her partner created, she observes the eclectic gathering of shrubs, trees and perennials that pay tribute to her travels. "When I look at my garden, I see all the places I've visited softly merge," she says.

Only on Labofem—a YouTube channel that she launched to share her deepening knowledge of botany with other Turkish green-thumbed enthusiasts—does the outside world intersect with Güçlütürk's new existence. Here, interested viewers seeking advice on their own plants can peruse videos where she counsels on everything from selecting the right planters to identifying common wintertime ailments, and watch short, snappily edited films that chronicle Güçlütürk's activities at home.

"YouTube is where I go to share my experiences," she explains. "The [viewers] don't judge me for my hair or makeup—which I don't have anyway—they just listen to what I say. I play the role of an entertainer and, while entertaining, I can help them look at their plants."

There are reciprocal benefits to this online community. Often, a question is posed that Güçlütürk must research to answer. "Both sides are winning: They listen and I listen," she says.

"I live in a vegetative state."

Güçlütürk rotates her selection of indoor plants according to the season ("If I don't limit myself, I would have much more," she says). She finds each one a suitable spot—like a high ledge for the fishbone cactus (*Epiphyllum anguliger*), pictured above right, since it is best displayed as a hanging plant.

In the winter, the temperature of the greenhouse drops and more of her "precious and small" plants make their way indoors.

When Güçlütürk began to study botany, she says she had to donate 1,000 novels to a secondhand bookstore in order to make more room for her plants. "Now, I have only books about plants," she says of the stacks of reference books on "local flora, deep botany and ecological mindfulness" that fill her shelves.

Güçlütürk takes a walk every day in the woods near her home and fills her notebook with illustrations of plants and flowers she sees along the way, looking them up in her books when she returns home. She also uses her notebook to keep track of the flowering patterns in her garden so that she can redesign and give the garden an "experimental new look" as certain plants fall dormant.

ASK ANKER AISTRUP

&

MAR VICENS

Some outdoor spaces require careful maintenance to thrive. Others—like the boulder-strewn Mallorcan grove where Ask Anker Aistrup & Mar Vicens sensitively restored The Olive Houses—are best cared for by being left entirely undisturbed.

High in the Tramuntana Mountains of Mallorca lie two tiny houses made from slabs of local stone. Set between millennia-old olive trees and a scattering of craggy boulders that jut up from the fertile ground, the houses are scarcely visible, which is precisely how owners Mar Vicens and Ask Anker Aistrup intended it.

"The project is reminiscent of a cave—man's first home," says Vicens. A Spanish architect whose grandfather built a home on Tramuntana when the area still lacked access to electricity and running water, Vicens spent her childhood clambering over the mountain's rocky landscape. For Danish-born Aistrup, accustomed to flat terrain, the dramatic topography was intriguingly unfamiliar. Before the duo embarked on construction, they decided on two rules: They would not cut down any olive trees, and they would not move the rocks.

Vicens and Aistrup cofounded the studio Mar Plus Ask in 2015. Their projects have spanned family villas in Valencia to converted industrial office space in Berlin, all of which are underpinned by a core tenet: designs intended to last. "We try to think of things we do on a scale of at least 100 years," says Aistrup, who points to the protected status of historic architecture across Europe as an example of the cherished role that buildings can assume if built with care. The Olive Houses, as they coined their Mallorca venture, were a response

to this proposition. How could they create something new that would live for years to come within an ancient landscape?

The Pink House—lit by a skylight and housing only a bed and a fireplace—was built around a large chunk of rock whose craggy gray bulk rests against soft pink walls. "The feeling with the natural skylight and the rock is that it makes you look at nature differently," says Aistrup. "It brings it into another setting."

A nearby toolshed was converted into the Purple House, in which a simple dining and kitchen space, as well as a small bathroom, can be found. A frameless window opens to a vista of more olive trees.

"When you're inside, you see the trees very clearly," says Aistrup, who explains that they deliberately selected the paint tones—light pink and deep purple—to complement the shades of green found on their leaves.

Since The Olive Houses were completed in 2019, Vicens and Aistrup— who are partners in life as well as business— retreat from their base in Valencia and stay there often with their young daughter. They have also offered it to other creative professionals in search of a quiet hillside on which to think. "I think people fall in love with it because it's raw and wild," says Vicens. "We come here from our terribly domesticated lives and can finally take a big breath of freedom."

The olive groves of the Tramuntana Mountain Range are a part of Mallorca's cultural heritage; for centuries, olive oil has been at the heart of the island's economy and life. The most common variety of tree in the area is the empeltre, which produces a black olive with a slightly higher-than-average level of acidity.

The couple poured the Pink House's
cement floor between the site's existing
rocks. Smooth and rough in equal measure, the design blurs the boundaries
between inside and out.

The dusky pink-and-purple palette that the couple selected for the property's stucco walls was chosen to complement the matte green underside of the olive leaves found in abundance around the buildings.

"We come here from our terribly domesticated lives and
can finally take a big breath of freedom."

To reach both houses, visitors are forced to meander and turn through the huge rock formations and ancient trunks of the landscape. The stone terrace walls and the couple's gentle trimming of the olive trees are the sole signs of human intervention.

UMBERTO PASTI

The vision for Rohana came to Umberto Pasti as he dozed under a fig tree twenty years ago: a garden filled with all the flowers endangered by Morocco's industrial sprawl. Today, the twenty-five-acre (10 ha) site has become a botanist's mecca.

Umberto Pasti believes the light of northern Morocco resembles that of Italian Renaissance paintings. "It's a marvelous light, but one that knows no mercy: its brightness is absolute," muses the writer and horticulturist, who splits his time between Milan and Tangier. "The light from a dream."

It was in a dream twenty years ago that Pasti envisioned what he calls his "combat garden": Rohuna, where he rescues wild flora threatened by Tangier's unbridled industrialization. That day, he had been looking for plants in the Moroccan countryside and had fallen asleep under a fig tree when the idea to rescue rare flora struck him. Pasti, who had inherited some money from his father, acted as if he were possessed, he says, buying the land, and promptly employing people to help. "Usually, making a garden is a long process, but in the case of Rohuna I already knew what I wanted to do."

But Pasti had done his homework, too. He had fallen in love at first sight with the coastal city of Tangier a decade earlier, during a holiday, after heading north to flee the socialite scene in Marrakech. Having missed the exit for the town, he arrived at a giant field of *Iris tingitana*, a bright blue flower native to the area, which is becoming increasingly rare. "Before the sea, there was this other sea of flowers," he

recalls. He decided to stay. The first home he bought there would include his first garden—a courtyard oasis, as is traditional in Morocco, which he replanted with fruit trees, roses from Italy and exotic plants from Asia and South America. By the time he started working on Rohuna, he had been studying the local flora for years and had landscaped over a dozen other gardens.

An early lover of nature, Pasti kept thirty frogs and a pet snake in his Milan apartment as a child. For him, caring for nature is akin to loving life. At Rohuna—twenty-five acres (10 ha) of land which he hopes to preserve as a foundation—he originally had one strict rule: Only rescued plants allowed. He is more flexible now, but the garden is still full of endangered, even functionally extinct species—like the pastel yellow *Iris juncea* var. *numidica*. Today, Pasti tries to focus on plants that do not need watering and advises fellow gardeners to do the same: "Plant seeds that will be autonomous after a year and be patient."

Rohuna, once a dream, has become a destination for botanists: Any donations help to fund recycling in the nearby village and support local schools and the children that will look after the plants in the future. "Nature is circular," Pasti says. "It's beautiful how the plants we rescued now help people."

The walkway outside Pasti's pavilion is dotted with potted plants. On the desktop of his studio, he has a large arrangement of budding calla lilies (*Zantedeschia aethiopica*). When Pasti is in Tangier, his head gardener rotates the flowers in each room twice weekly.

Because his three eucalyptus trees dry out the soil surrounding the pool, Pasti uses potted plants to bring greenery to the area. In his sitting room, he likes simple plants "like you would see in the neglected courtyard of a Roman palazzo," such as the spider plant (*Chlorophytum comosum*) pictured on the column fragment.

Large parts of the garden are covered in fast-growing Australian ivy (*Muehlenbeckia complexa*), pictured above right. The Chinese vase in the sitting room contains an arrangement of Moroccan irises (*Iris tingitana*), while the pot in the foreground contains a hybrid hellebore (*Helleborus orientalis*) that flowers perennially.

Pasti's tropical hydrangea (*Dombeya x cayeuxii*), pictured above left, flowers in the winter and smells like honey. Outside the pavilion, pictured opposite, he's sunk a shallow basin and filled it with yellow flags (*Iris pseudacorus*) rescued from a pond that was "swallowed up by a new road" nearby.

LINDA TAALMAN

The Californian desert wouldn't be many people's first pick for building a house reliant on nature, but architect Linda Taalman has proved otherwise: Her self-sufficient glass home is a testament to the power of thoughtful design and careful land stewardship.

The idea for Linda Taalman's house was born long before the site was found. The California-based architect's vision was of an airy structure set in a serene environment. She had struggled to find the right spot for her new home in Los Angeles, so she looked for her ideal location outside of the city. "I'd been spending a lot of time in the desert, and it made sense to me that I should build it there, where I could really maximize the benefits of having a glass house that could have a relationship to the landscape," she says.

The desert environment is, of course, not known for lush greenery, so there was little opportunity for traditional gardening. Nonetheless, the environment is key to the house's character. Although Taalman had to cut into the site's rocky slopes so that the house could sit flat and accommodate the patio, she did it in a way that looks seamless by incorporating the plot's natural features into the design. The effect was enhanced by floor-to-ceiling panoramic windows in the kitchen and living room.

In keeping with Taalman's desire to be at one with nature, the home makes as little environmental impact as possible—its light form is made of prefabricated glass, steel and aluminum components that slotted together with minimal

building work on-site. "It's really a kit of parts that was precut and predrilled and assembled," Taalman says of the process.

The house is also entirely self-sufficient and provides for its own energy needs through solar panels, and by taking advantage of how the natural light moves through the building. In the winter months, the home is heated by the sun during the day and heat-storing floors at night; in the summer, large glass sliding doors open out onto two courtyards, keeping the whole building cool and airy. Shade is provided by an overhanging roof and by the placement of the solar panels. The reflective aluminum and steel pick up colors from the surrounding land and sky, which change with the seasons.

"You feel immersed. You have this view of animals, birds, clouds, the sun, moon or stars. At night you can turn all the lights off and have an amazing celestial experience," Taalman says. "Living in a space like this gives you a heightened awareness of your environment. There's a sense of being as close as you can possibly be to the landscape, but at the same time having the protection of a building that allows you to stay there and contemplate it rather than escape when it becomes uncomfortable."

The scrub oaks and juniper tree offer a scenic view from inside the bedroom but also afford the room privacy from the road and help to shelter the house from the wind and the elements. "All of the plants are incredibly slow growing, and each has its own hearty beauty from surviving in a landscape with so little water," says Taalman.

One of Taalman's favorite plants is the creosote bush (*Larrea tridentata*, pictured opposite)—a humble-looking but highly fragrant plant. "When it rains, it makes the whole desert smell intoxicating," she says.

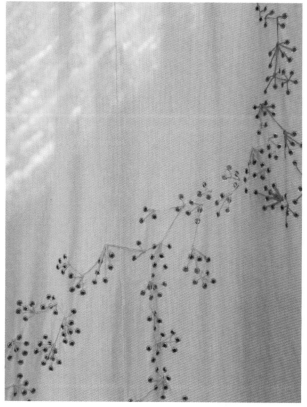

"Living in a space like this gives you a heightened awareness of your environment. There's a sense of being as close as you can possibly be to the landscape."

YASUYO HARVEY

For Yasuyo Harvey, caring for nature means observing its small quirks. Her south London house reveals a strange symphony of plant life: walls of pressed leaves, sculptural assemblages of dried branches, and an ever-changing display of cut flowers.

For Yasuyo Harvey, home is about getting back to basics. From the outside, her 1930s house in southwest London appears typical of others in the neighborhood. Its interior, however, is a minimalist meditation with floral accents: White walls, wooden surfaces and raw plaster are offset by ceramics brimming with dahlias, magnolias and Japanese anemones, along with boughs of larch, Norway maple and flowering cherry. Harvey is a botanical artist, and her creations seem informed as much by an underlying Japanese sensibility as they are by nature itself.

Harvey has loved flowers ever since she was a girl growing up in Kyoto, where she practiced ikebana as a hobby before moving to London to study in 2004. She found work with a florist in Piccadilly Circus and did flower decorations for corporate events, weddings and other occasions, and later gave floral art classes to Japanese expats. She began going to Paris Fashion Week and eventually met and then collaborated with designers Faye and Erica Toogood, which led to more floral styling for photo shoots and interior design projects.

"My passion is to make sculptures with botanical elements. I can express myself," says Harvey, pointing to an arrangement of skeletal leaves coated in liquid latex in the fireplace of her living room. "I think about why we decorate homes with dry flowers. In a sense, it's similar to mummification. So, one thing I've been interested in is capturing plant decay in my sculptures."

Harvey's home, which she shares with her husband, Phil, and son, Noah, is full of delicate dried plants, ferns, grasses, seedpods, seaweed and coral. Rather than gaudy blooms, she prefers subdued, structured objects, which she gathers from her garden, while on walks or from the wholesale flower shops she frequents. She molds these into sculptures, collages and assemblages that act as reminders of the cycle of growth and decay. In that sense, they resonate with themes of nature and ephemerality in traditional Japanese aesthetics, but Harvey is reluctant to use terms like *wabi-sabi* to describe her work.

"As a term, *wabi-sabi* is overused—it's become a buzzword. But as I was born and raised in Japan, it did have a fundamental influence on me," says Harvey, who counts ikebana and bonsai, but also the Dutch garden designer Piet Oudolf, among her creative influences. "Maybe *wabi-sabi* helped a lot in deciding the details of my home, in addition to combining old and new. I think I'm good at mixing different elements."

Harvey hopes her botanical artworks inspire others to notice their layers of detail, as well as the outside world in general. "I don't have a full-fledged philosophy per se," she says. "But I do notice plants and other small things in the natural world that others may overlook." Even going for a walk in the woods can inspire people to take an interest in and care for the flowers, trees and other plants, she suggests.

The shelving unit pictured opposite is used to store the dried objects Harvey creates or collects. On the bottom shelf, corkscrew rush (*Juncus effusus*) is arranged in a *kokedama* moss ball. An overgrown *Lepismium* succulent dangles from the top shelf. Pictured above, a purple hellebore from Harvey's garden.

Above, Harvey arranges hellebore with mimosa (*Acacia baileyana*). Opposite, she waters a winter aconite (*Eranthis hyemalis*) planted with moss in a stone container. "It was inspired by Japanese winter gardens," she explains of the plants she selected. The aconite is a substitute for the Japanese fukujyusou flower, which is hard to find in the UK.

The rabbit's foot fern (*Humata tyermanii*), pictured opposite, gets its colloquial name from the fuzzy rhizomes that grow around it. The rough horsetail (*Equisetum hyemale*), pictured above, is a popular element in ikebana arrangements and a common riverside plant in many parts of the world.

JONAS

BJERRE-POULSEN

When architect Jonas Bjerre-Poulsen bought a 100-year-old house in a Danish fishing village, the garden presented a puzzle: How could it be modernized in line with the family's needs, and Bjerre-Poulsen's own minimalist leanings, without losing its charm?

When describing his own work at the Copenhagen-based studio Norm Architects, founder Jonas Bjerre-Poulsen uses words like "minimalism," "simplicity" and "restraint." It might, then, come as a surprise that the garden of his own home on the outskirts of the Danish capital is a rambling, multifaceted plot that is characterized by wildness, romance and discovery.

The house itself was built in 1911 and is located in Vedbæk, a fishing village on the Danish coast about a half-hour train ride from the center of Copenhagen. "When it was built, the architectural style of southern Germany and northern Italy was very much in vogue. It looks like a Tyrolean [Alpine] house," Bjerre-Poulsen says. "It's perhaps a bit odd compared to the fishermens' houses around it."

When Bjerre-Poulsen and his family took over the property, a series of modifications over the decades had left it looking somewhat tired. The garden had also been neglected—the reflecting pool, for example, was covered in dirt. They began renovating the house and garden to restore its original character. "As an architect, you always need to consider the spirit of the site," Bjerre-Poulsen explains. "I've designed the garden to keep with the character of the house."

The garden today is a symphony of reflecting pools, climbing roses, sumptuous flowerbeds and ornate court-yards, which all sit in harmony with the home's traditional architecture. The reflecting pool retains its original form and granite border stones, but is now heated so that it can be used as a bathing pool for Bjerre-Poulsen's children. "It's like a Jacuzzi, but it was important that it didn't look like it," he says.

In the planting and configuration, Bjerre-Poulsen brought in outside influences he encountered in his travels and work. "I'm very inspired by Japanese formal gardens, so it's a bit of a hybrid. Some areas have pebble stones and cherry trees." It was important to him that there would be spaces for everyday outdoor living, so he added scattered clusters of tables and chairs. "If a table is there, you'll use it. If you have to bring it outside, you never will," he says. Meanwhile, the wooden shed, once a pavilion, has been transformed into a studio and library, complete with blackout curtains so Bjerre-Poulsen can use it as a photography studio.

Between these areas of human activity, the garden remains paramount. "Crossing the garden while walking from the house to my studio is an integral element of the design. There is no other way [to access it] so you have to encounter the climate and elements when moving between the spaces," he says. "It's a joy to have that connection to the outside as part of your daily routine."

Bjerre-Poulsen's garden beds include bay laurel (*Laurus nobilis*), hydrangea, catmint (*Nepeta*), pink roses, grasses and an olive tree. The leaves of an indoor camphor tree (*Cinnamomum camphora*) peek out from his photography studio.

The sunken pool in Bjerre-Poulsen's grass lawn—transformed into a heated bathing pool for his children—is afforded privacy by shaped boxwood. Right, catmint and laurel sprout through the paving stones. "The plants only get to stay if they can survive on their own," he says.

ABDERRAZAK BENCHAÂBANE

Abderrazak Benchaâbane gave Marrakech its lungs: He created the city's expansive Palmeraie Museum and restored Yves Saint Laurent's celebrated Jardin Majorelle. It is in his own garden on the city's outskirts, however, where he creates his "visions of Eden."

For Abderrazak Benchaâbane, the vegetation of Marrakech is a relief: Plants signal an oasis, the presence of life in an otherwise hostile environment. "I feel protected in gardens," he says. "Outside feels a little dangerous."

The sixty-one-year-old garden designer—also a museum owner and perfumer—has devoted his life to ethnobotany, the study of plants and the people who live alongside them. "I've tried to bring gardens with a history back to life, to grasp their spirit without betraying their author, even though gardens are artworks that escape the control of their maker," he says. Benchaâbane's best-known project is his restoration of the Jardin Majorelle, a cobalt-blue bolt-hole of rare plants in the heart of the Moroccan capital, which he undertook on behalf of the late Yves Saint Laurent and his partner, Pierre Bergé. He spent a decade reviving the garden as it was initially envisioned by French orientalist painter Jacques Majorelle in the early 1900s, combing through correspondence, journals and archives to retrace its genesis.

"You start digging the soil, and the soil starts speaking," says Benchaâbane, comparing his work to that of an archaeologist. "For instance, the garden was named *bou safsaf*, meaning 'where the poplars grow,' in the official register, but there were no poplars in the garden. Still, during the works, we dug out poplar trunks." He left Majorelle a few months before Saint Laurent died in 2008, having restored its waterways and breathed new life into the garden, now famous for its abundant flora, including giant cacti, palm trees and bougainvillea.

Benchaâbane's passion for nature dates back to his childhood: His father's family were farmers who taught him how to care for his surroundings at a young age. His mother came from a lineage of palace and hammam (communal bathhouse) builders. He inherited her love for Moroccan art and sense of the mutually beneficial relationship between humans and plants. "She nursed me with herbal remedies, and I quickly realized that people didn't care for plants, but plants cared for people," he says. The garden he cherishes the most is the one he restored as a gift to Marrakech's Al-Antaqui hospital, where he was cured from a life-threatening illness as a child.

Young gardeners should know two things, Benchaâbane says. First of all, that each garden is unique, much how children from the same parents have different characteristics. "They are a vision of Eden as imagined by each person," he insists. Gardeners should also realize they are not working alone. "Time is an invisible yet unavoidable partner. There's no point in being in a rush."

Today, as the founder of Marrakech's Palmeraie Museum, a contemporary art gallery where visitors can stroll through five acres (2 ha) of wet and dry gardens, he hosts recycling workshops for children and promotes sustainability. His own home, and garden, are next door. He says hundreds of birds have started to nest there over the years. "It's a good sign. Birds nest where there is love."

Benchaâbane started the Musée de la Palmeraie to house his growing collection of contemporary Moroccan art. Located on a 150-year-old restored farm, the two-and-a-half-acre (1 ha) expanse includes formal, walled gardens such as the Andalusian Garden, pictured left, and the ethnobotanist's private home, pictured above.

A fragrant rose garden, pictured above,
inspires Benchaâbane's projects as a
perfumer. In 2001, Yves Saint Laurent
asked him to develop a perfume that
encapsulated the Majorelle Gardens.
Its success eventually led to the launch of
Benchaâbane, his own line of fragrances
inspired by Marrakech.

A shaded seating area decorated with traditional Moorish tiles is hidden within a Mediterranean-style aquatic area of the Andalusian Garden. Visitors are encouraged to relax in the shade and watch koi carp circle the water pool, pictured opposite.

The Cactus Garden, pictured opposite, was planted a decade ago and features over forty varieties of cactus from Morocco, South Africa, the USA, South America and Mexico. Some of the palm trees on the site are over a century old.

LOTUSLAND

How did a six-time divorcée with a failed opera career go on to create Santa Barbara's most beautiful garden? The genesis of *Lotusland* is a story of ambition, spirituality and the "dramatic flair" of its singular founder.

Ganna Walska never wanted to be the steward of a large garden. She wanted to be an opera singer. But despite a flamboyant presence, striking physical beauty and a charming insouciance that many performers would kill for, she had a short-lived career. Private observers claimed she possessed considerable talent, but she suffered from what the *New York Times* in 1929 called "a pernicious form of stage fright," rendering her instrument powerless on stage. Fortunately—for her and posterity—the Polish socialite was able to shift her desires to more accurately correspond to her natural talent for horticulture.

When Walska was in her fifties and on to her sixth marriage, she looked to California as a fresh start. In 1941, she bought a thirty-seven-acre (15 ha) Santa Barbara estate at the urging of her husband, who envisioned the land as a retreat for Tibetan monks. She lovingly transformed the property, christened Lotusland in a nod to her eclectic spirituality, into one of the world's largest and most diverse gardens—home to more than 3,300 plant species from every continent. Walska devoted the rest of her life to developing and maintaining Lotusland, as well as ensuring its legacy for continued generations to come.

Walska's devotion is carried on by her public charity and its staff, who go to great lengths to uphold her vision. According to Paul Mills, the curator of living collections at Lotusland, Walska channeled her artistic energy into the garden, learning about plants and their care, and arranging them to express her creative vision. She applied what Mills calls "dramatic flair" to her gardens: "Her modus operandi was mass planting," he says. "If one plant is good, 100 is better."

What's more, "she believed in the spirit of plants," says Mills. "It was hard for gardeners to get things done. The bromeliads, for example, were very prolific but she wouldn't let them throw anything away—which you need to do sometimes. They had to create a whole other bromeliad garden."

Her vision resulted in a diverse, highly distinguished set of gardens that are at once playful and irreverent. The monochrome Blue Garden is populated solely by grasses, trees, succulents, and bushes with silvery-gray leaves, like an oddly temperate winter fairy tale. Elsewhere, Lotusland's Cycad Garden is home to three *Encephalartos woodii* plants referred to as The Three Bachelors. Extinct in nature, these stalwarts are are among the last remaining members of their species. All males, they dutifully produce cones year after year that will never have a female equivalent to fertilize.

To protect these precious specimens from potentially fatal fungus, Mills and his staff have undertaken a years-long project of removing the plants by the roots, painstakingly sifting any fungus-bearing wood elements from the soil, then replacing the cycads in their original formation.

Like many things in nature, the relationship is symbiotic: "The plant collections here keep me going," says Mills.

There are six different cacti species in the garden surrounding Lotusland's pavilion building, pictured opposite. Staff estimate that the clustered globular golden barrel cacti (*Echinocactus grusonii*) are 100 years old. The large green plants that skim the pavilion's gables are not cacti but candelabra trees (*Euphorbia ingens*).

Ganna Walska chose to live in Lotusland's pink pavilion building, rather than the estate's main residence. The building is surrounded by a garden of bromeliads that brush up against its walls, including the red inflorescences (*Aechmea weilbachii*) pictured below.

At the entrance to Lotusland's tropical garden stand two aromatic Tasmanian blue gum (*Eucalyptus globulus*) trees, shown above. Mills estimates that the trees are around seventy years old—old enough that their branches have grafted together through a natural process known as "inosculation."

ALEJANDRO STICOTTI

&

MERCEDES HERNÁEZ

Most gardens are built around houses, but some houses are built around gardens. Alejandro Sticotti's unusual design project began when he bought an overgrown plot of land in Buenos Aires and decided to sink his home into its junglelike vegetation.

The home of Argentine architect Alejandro Sticotti and graphic designer Mercedes Hernáez is a transparent cube sunk into a lush garden. It reflects Sticotti's signature style: clean lines, wood on the walls, and glass panels that blur the boundary between outdoor and in. The structure was conceived of when Sticotti was looking for a house to buy in Olivos, a suburb of Buenos Aires, but then decided to take a less conventional approach; he chose the overgrown garden in between two residences instead.

It took three years to build the house, which is covered in slatted wood and anchored with polished cement floors. Because it was a personal project, Sticotti crafted each element to his liking. "I made the living room with double height [ceilings]—not everyone likes that," he says. The house, like his studio in the city's Palermo district, is cloaked in green. Banana trees and vines hug the structure; afternoon light rushes in as the clouds shift and the windows are large enough to watch the sun rise and set. He says, "I like the sensation of being outside when you are actually indoors. I don't like closed spaces."

The arc of light that comes with the sun's movement captures nature's daily transitions. "The house is constructed in such a way that the sun will rise and move

around it," says Sticotti. The overall effect is of an inverted greenhouse, a framed glass structure ensconced by plant life. The garden is teeming with native and nonnative plants, like a large *Jacaranda mimosifolia* tree that Sticotti planted; it drops electric indigo flowers during late spring, littering the yard with color. The ever-growing collection of succulents is Sticotti's favorite feature. There are no plants in the house, however. He prefers the greenery as a theatrical backdrop.

Weekends are spent in the garden, trimming and weeding; the couple does all of the maintenance themselves. "We mostly work in the garden and we love to cook," says Sticotti. The herbs growing there—a snatch of thyme, a few leaves of laurel—find a way into their meals. There is a traditional grill, a *parilla*, set just outdoors but in close range to the kitchen inside. In the early evening, the couple starts a fire to grill an array of meats, served with good Argentine wine, and pasta fragrant with the basil grown within arm's reach.

Now, the house has aged and the cacti have risen above head height; it is a life lived among the plants. Is there anything Sticotti would like to modify about the garden? "I wouldn't change it for anything," he says. "This is the way I like it."

In the summer, the couple spend most of their time on the decked terrace that overlooks the garden (pictured left). "In the afternoon, after watering the plants, we usually collect all the dried leaves and make a small fire," says Sticotti. "We also usually have lunch there on Saturdays."

The couple designed the garden to look untamed. A purple flowering crape myrtle (*Lagerstroemia indica*) tree shades the terrace, pictured opposite, and succulent ghost plants (*Sedum weinbergii*) hang over the window of the master bedroom, pictured below. "It has no defined beds," says Sticotti. "We like it to look wild."

"I wouldn't change it for anything.
This is the way I like it."

CAMILLE MULLER

Whether in Madagascar or the Ionian Islands, Camille Muller approaches each garden design as a lifelong commitment to care. But none hold the same place in his heart as his own secret garden on a Parisian rooftop.

"This is not a thought-through garden like the ones I do for my clients," says landscape gardener Camille Muller as he climbs up the ladder to his Parisian rooftop garden. Above Muller's 11th arrondissement home, an oasis of climbing rose bushes, blue-purple clematis and towering fruit trees unfold. Were it not for the faint sound of honking horns in the distance and the occasional zinc roof panel visible underneath, one might think the site was in the middle of the countryside. "I wanted to create a place of refuge in the city, where nature is left to take its course," he explains. "Many of the plants have been brought here by the birds or the wind. Others, I found abandoned."

The son of an agricultural engineer, Muller owes much of his green thumb to his upbringing. His youth was spent in the rolling foothills of the Vosges mountains in eastern France, between the family home (with its organically composted vegetable garden), the local forests and the agricultural school that he attended. "I learned everything on the ground," he says. "Gardens have always been my dreamscapes, places where I could express myself freely." By the age of thirteen, Muller knew he wanted to make a profession out of creating his own "dreamscapes," rooting his practice in the ecological beliefs and intuitive approach of his early years.

Since his first commissions in the 1980s, Muller has worked with private clients on everything from green rooftop terraces in dense urban settings to large-scale rural gardens and sprawling orchards. All of his work comes from a respect for the local habitat—the "biotope," as Muller calls it—which involves adapting the garden's vegetation to the soil, climate and topography, and makes chemical products redundant. "And yet," says Muller, "my work is about striking a balance between the client's demands and the needs of nature." For a project located in a dry area of Madagascar, Muller created a fake humid biotope, installing water misters and shipping in traveler's palms to allow his hotelier client to get the tropical environment he wanted. A couple of decades later, the place has become an ecosystem of its own, abundant with exotic flowers and birds.

Muller regularly revisits his gardens to make sure they are well maintained and that the plants have what they need to survive. One of the sites he enjoys going back to the most is in the Greek Ionian Islands, where he transformed a seventeen-acre (7 ha) cliff-fringed peninsula into a terraced Mediterranean landscape dotted with fragrant herb bushes, olive and cypress trees. At once wild and manmade, it is perhaps closest in approach to his own Parisian rooftop garden. "Intuition is what gives my projects soul and makes them meaningful beyond the aesthetic dimension," he explains.

Muller's rooftop garden is home to many
potted plants that require pruning. The
Cape African-queen (*Anisodontea capen-
sis*), pictured opposite, for instance, needs
a light trim after flowering to encourage it
to do so profusely and repeatedly.

"I wanted to create a place of refuge in the city, where nature is left to take its course."

In winter, the automatic sprinkler is turned off and Muller uses a hose or can to water everything from the Narihira bamboos (*Semiarundinaria fastuosa*) and 'Glacier' ivy (*Hedera helix*) to the holly oak (*Quercus ilex*) planted on the rooftop and the potted Jerusalem sage (*Phlomis fruticosa*) and weeping sedge (*Carex pendula*) on the terrace. "These plants often die more from thirst than from cold in winter," he says.

Five Tips

Houseplants are low-maintenance guests. They breathe life into your home and require very little hospitality in return; just a comfortable spot to rest and a little water. The following pages contain some sage advice on the basics of houseplant care and correct some popular myths: like why gravel isn't that great a drainage solution, and why sunlight and sky light are two very different propositions for your plants. Follow these tips and you may just keep that *Monstera* alive a little longer than the last one.

How to Care for Houseplants

Words by Darryl Cheng

Founder of *House Plant Journal* and author of *The New Plant Parent*

A plant's pot should not be chosen with appearance in mind, but rather the plant's fundamental needs. A large plastic nursery pot that has adequate drainage holes is the best home for a houseplant. You can set this nursery pot inside a larger, decorative pot—after watering, wait an hour or so for it to drain and then pour away any excess water.

If you must put a plant into a pot without drainage holes, do not (as is often advised) add a layer of gravel as a drainage layer; this only creates a space where root rot bacteria can fester. Instead, just be cautious when watering: Ensure that you pour slowly and that the volume of water is no more than a quarter of the total amount of soil. Water again when the first inch (2.5 cm) of soil is dry.

The indication that a houseplant has outgrown its pot is when the roots have coiled up at the base of the pot; this is a warning that the plant has become "rootbound." Unpot the plant and gently tease apart the roots while removing about half of the old soil. Take the opportunity to cut off any brown and mushy roots that may have rotted; healthy roots are firm and off-white. Transfer the plant to a new pot that is no more than two to four inches (5 to 10 cm) larger in diameter than the old one.

2. HOW TO FIND THE RIGHT LIGHT

People tend to put houseplants where they think they will look nice, rather than where they will grow optimally. But light is essential. It drives photosynthesis—the plant's ability to make its own food. Most houseplant care instructions call for "bright indirect light" but offer little guidance on how to find it. And, although it can be easy to find a spot of direct sunlight, it can be difficult to gauge the level of light because the human eye adjusts to all levels of brightness.

If you want to give your houseplant the best chance at "bright indirect light," put your plant in the place that has the widest possible view of the sky. And, if the sun will shine directly on the plant for longer than two or three hours a day, shade the window with a sheer curtain. All indoor plants will want to see as much of the sky as possible, but only some will actually need to spend a few hours in direct sunlight each day.

3. HOW TO CLEAN & PRUNE

Clean

Plants with larger leaves that live indoors will inevitably collect dust. Every couple of months, you can simply wipe them with a damp cloth or the inner side of a banana peel. If the plant is not too cumbersome to move, you can also take it to the bathroom and shower the plant with room-temperature water. Do not use any soap, though, as the detergent harms the waxy coating of most tropical houseplants.

Prune

It is important to remove and discard dead leaves. They won't decompose properly indoors since there are—hopefully—few insects to perform the task. Pruning generally applies to branching plants, and the regularity with which you need to do so will vary from plant to plant. But it is also a matter of aesthetics and can be done when a plant begins to dominate its surroundings. Cut a stem off at its growth point, and another will grow back stronger.

If a friend or family member can't help with watering your plants while you travel, then you should consider the three types of watering patterns that your houseplants will need in your absence. In general, the more light the plant receives, the faster the water usage. So, if light levels decrease, then the soil will not dry as quickly. Houseplants that require evenly moist soil regularly will be the most difficult to keep hydrated while you're away. Move these plants somewhere that will never see direct sunlight but that still has a wide view of the sky. Assuming the pots have drainage holes, you can keep them submerged in a shallow pool of water. Give houseplants that require water only when the soil is partially dry a thorough watering before you leave and move them farther from the window. For plants that need water only when completely dry, simply move them away from any direct sunlight and forget about them until you get home.

There is a common misconception that houseplants stop growing in the winter, but consider the indoor temperature: It does not vary as widely as it does outdoors. Although there may be fewer hours of daylight, the sun, which is lower in the sky and unobstructed by usually leafy trees, may penetrate deeper into a room. Instead of thinking there is some different regiment of care during winter, simply practice responsive, observational houseplant care.

Water by observing soil dryness: When the soil reaches the plant's preferred point of dryness, water it. Fertilize when you see active growth on the plant—let the plant's active growth tell you when to fertilize, not the outdoor season. In places where there is significantly less light in the winter, move houseplants to a warmer room—most will appreciate a temperature of 59°F/15°C. Do not place houseplants next to a radiator, however. This can quite literally cook them.

Gardening
Tools

Protect your clothes and stay warm with durable outerwear. Finnish brand Fiskars teamed up with fashion designer Maria Korkeila to create modern staples for the garden, including the brightly colored gilet, hat and neckerchief pictured right.

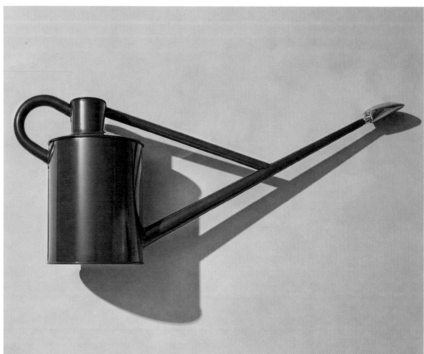

Small plants, pots and seedlings will not survive without a watering regimen. Watering cans, like the one pictured right from British brand Haws, are good for plants and better for the environment than a sprinkling system or hose pipe.

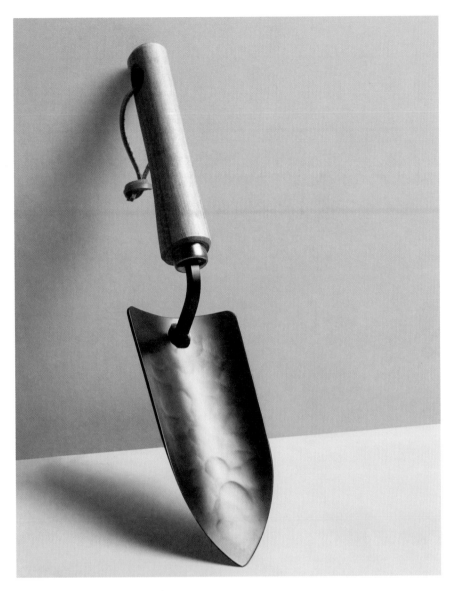

Good tools do not make a good gardener, but they do add a little pleasure. This trowel—useful for greenhouses, conservatories or light garden use—was forged for Japanese brand Niwaki.

Gardening scissors are neccessary for cutting flowers, deadheading and light pruning. The pair above, from Niwaki, are strong but delicate for accurate cuts.

Use secateurs instead of garden scissors for heavy-duty pruning. A clean cut prevents disease, so a sharp pair—like the pair above from Niwaki—is ideal.

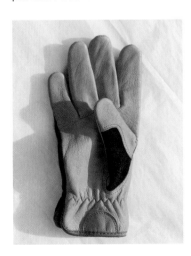

Getting your hands dirty is part of the fun, but a good pair of garden gloves, like those above from Hestra, can keep hands protected from thorny plants.

Iacobo Kempener pix

Io. Theodor de bry fcalp.

PART TWO

Creativity

The laws of nature have long inspired the principles of design.
Meet the creative professionals finding new ways to interpret
the abundance around them.

MAURICE HARRIS

As a child, Maurice Harris was inspired by the floral arrangements his grandmother made for the church pulpit. Today, he's continuing her legacy—and building his own—by creating extravagant arrangements for communities near and far.

Maurice Harris didn't exactly plan on becoming a florist, but his decision to express his creative voice through flowers was strategic. "I was always figuring out ways to be creative, whether through hair, fashion or makeup," he says from his office in Los Angeles. "But flowers were a space that wasn't heavily saturated with black people, black men, black gay men. Those other industries have their fair share of stars. I felt this was the industry where there was the opportunity to breathe new air."

Under the brand name Bloom and Plume, Harris's rococo creations are indeed light-years away from the baby's breath and pastel roses of traditional flower arrangements. On Instagram, his captions are laden with the sort of slang and Internet memes that would make the cardigan set clutch their collective pearls. Social media has been a way for Harris to invite more people to participate in a world that has hitherto been occupied by the (white) ladies who lunch.

His has been a popular invitation. He has over 150,000 followers and growing. "My platform is very interesting," says Harris. "Most of my clients don't look like me, but my Instagram following is very diverse."

Photographs are useful as a visual aid, Harris believes, but they don't adequately capture the experience of engaging with his arrangements. A new café he's launched in LA's Filipinotown will feature his arrangements on each table. He hopes the café will make his art accessible in person to people who've perhaps never seen a high-end flower arrangement. "You get to come and be a part of our brand and community," he says. "You get to feel really seen."

Harris's arrangements take cues from his family's history. Flowers were a big part of his childhood community's celebrations, both big and small. At church, "my grandmother was in charge of doing the flowers at the pulpit. It was kind of a big deal, and she took a lot of pride in that." She also made the arrangements for the weddings of her eight children, photographs that Harris studied as a child.

"My grandmother had this design principle that I constantly use: a triangle, which for her represented the Holy Trinity," Harris says. Asymmetrical triangles are indeed very present in his designs, often in twos or threes.

The angular geometry of the bird-of-paradise flower, the floral emblem of the city of Los Angeles, also captures Harris's imagination. "It represents blackness for me, in a way," Harris enthuses. "It stands out above the crowd. It blends in before it blooms; it's got a weird nose or beak. And that beautiful vibrant blue! That orange!"

Harris likes to experiment with unusual blooms, such as the toffee-colored rose he is holding in the image above. The sepia-toned bloom was developed relatively recently by an Ecuadorian breeder, who cross-bred different species of rose to achieve the desired coloring.

BLOOM & PLUME

1640

BLOOM & PLUME
Studio
by Appointment

MAIL

Harris's outdoor workstation is crowded by plants on all sides, with an emphasis on red blooms such as the anthurium in the foreground and hanging lobster claw (*Heliconia*) behind the chairs. A bouquet of fringed tulips and a vase of red anemones sits on the indoor table, opposite.

LISA MILBERG

&

LEO FORSSELL

A vase of flowers is often considered a finishing touch, but not by interior designers Leo Forssell and Lisa Milberg. At their Swedish studio Arranging Things, projects only start after the pair has decided which plants will take center stage.

Arranging Things, the creative practice of Swedes Lisa Milberg and Leo Forssell, started organically and unpredictably. The two began as friends who shared eclectic design tastes and liked to go vintage shopping together. They then launched a website and Instagram account showing their finds, and now find themselves heading up an "umbrella organization of sorts" that combines interior decoration, retail, consulting and music curation (Milberg was formerly part of indie-pop outfit The Concretes, and is currently part of the band Miljon). They run a shop, but they do it on their terms: The storefront doubles as their design studio and is only open to walk-ins one day a week, on Thursdays.

Plants aren't explicitly in the duo's purview, but they are present in everything they do. Milberg and Forssell say that plants are often the first thing they budget for when they start working on a new brief, and, along with music, are the first thing to change in the shop or any other space when they want to create a mood. Their essential mission, as they see it, is antiminimalist, and they wouldn't be able to achieve it without plants and flowers.

"More than anything else, when we started, what we wanted to change was that a lot of interiors were very sleek and not very homey or personal," Milberg says. "They don't tell you much about the person in the space. We wanted to do something much warmer, friendlier and more welcoming. Plants are an essential part of that."

When it comes to vases, the two are drawn to organic or playfully shaped vessels that bear the marks of having been molded by hand. To fill them, they then look for plant life with the right proportions—although "right" can mean wrong. They favor arrangements with elements of surprise and juxtaposition, and which feature unexpectedly sculptural forms like the garlic flower. Usually, they'll curate a few complementary species together, with a theme in mind, but sometimes they like to go all-in on one flower: the tulip.

"We love tulips and one reason is their folksy quality, which they somehow manage to combine with an air of elegance," said Milberg. "They also look better and better with age—sometimes they look great into their death. The tulip is a common motif for David Hockney, and any motif that's good enough for him is good enough for us."

With sustainability increasingly at the front of people's minds, the duo has in recent years tempered their use of cut flowers and focused more on twigs, thistles and other cuttings that look good dried. When the flowers are fresh, they like to make them last by not only changing the water often but also by refraining from throwing out the whole bouquet at once. Different species have different life spans, so they discard dying flowers and put the survivors into a smaller arrangement, in a new vase. "Sometimes the second or third reincarnation of a bouquet is when it gets really interesting," says Milberg.

Above, from left to right, Forssell and Milberg have arranged red kale, dried *Stillingia* and Christmas roses (*Helleborus niger*) in vases by Anne Harström and yellow butcher's broom (*Ruscus aculeatus*) in vintage baskets. Hare's-foot clover (*Trifolium arvense*) sits in a tribute Matisse vessel. "It's the best piece of fan art we've seen in a long time," says Milberg.

In a corner of the studio, pictured opposite, preserved eucalyptus fills a mouth-blown glass vase by Aki Keitel. The Apila stool underneath—a vintage Lisa Johansson Pape piece—is shaped like a four-leaf clover. "*Apila* means clover in Finnish," says Milberg. "There's a matching table, too, but we've only ever seen it in pictures."

"Flowers don't always have to be real," says Milberg. "The tulip lamp [pictured above, at far right] is vintage from the '80s. There was also a daffodil version made. One day we'll track it down!"

JULIUS

VÆRNES IVERSEN

Tableau doesn't look like most flower shops. The Copenhagen studio, founded by Julius Værnes Iversen, more closely resembles an industrial jungle, where delicate blooms and brute materials are brought together in sharp relief.

There's an unspoken assumption that flowers are synonymous with nature. But not in the world of Julius Værnes Iversen. The Danish floral designer is the founder of Tableau, a studio that creates installations that juxtapose flowers with industrial materials, where the blooms themselves can end up looking alien and unfamiliar.

Often, that element of surprise comes from playing with vivid artificial hues, as with a tumbling mass of purple-painted asparagus fern that Tableau installed in their studio as "Christmas decorations" one year, or from suspending the flowers in midair like clouds. A hard-edged vessel, plinth or backdrop, often in steel or concrete, is the finishing touch.

"What I really like is the contrast between the natural and unnatural," says Værnes Iversen. "I think it gives a very contemporary vibe to a floral installation."

Værnes Iversen doesn't just revel in the look and the tactility of this approach; he sees it as a vital stepping-stone to floristry being recognized as a true form of art or design. "Right now, florals need to be put in a very unnatural stage to read as contemporary or conceptual," he says. Also key is Tableau's David Thulstrup–designed studio space in Copenhagen, which it uses as a gallery to host botanical-inflected exhibitions conceived with collaborating artists, enhancing its art-world credentials.

Værnes Iversen believes the ability to approach flowers creatively is something innate: "You don't need an education in this line of work. I believe that people should just jump into it and give it a go," he says. But he also respects traditional floristry, having spent his childhood working in his father's flower shops, which he now runs alongside Tableau. "It's such a big part of my soul. I feel that florals or botanicals have some sort of healing way with humans. If I have a bad day or if I'm ill, it's like my work is giving me energy. Do you know the feeling of having grass between your toes? I sort of feel like that when I work with flowers."

Tableau also makes some of its own products, which is an area where Værnes Iversen tries to be more sustainably minded. He uses shop-bought ready-made materials such as tiles and pipes and avoids manufacturing his vases from scratch, as a way to compensate for the adverse environmental impact of using cut flowers. He is honest about how problematic they are, believing there's "no sustainable way" of using them, but that they bring people joy and have value as an artistic object.

In his own home, Værnes Iversen applies a "more-the-merrier" approach to his potted plants, and offers the advice that plants will do best if you look at them and even talk to them every day. "Here in Denmark, it's very normal to say that you need to address your plants. I know it sounds like I'm maybe a bit of a nutcase, but I think they actually can track energy, and it helps, when you water it if you say, like, 'Wow, you look nice,' or whatever. I think it works."

On a shelf system designed specifically for his studio, Værnes Iversen displays sculptures and plants alongside one another. The tubular bulb plants (*Fockea edulis*) pictured are believed to be at least fifty years old. He sometimes puts flowers in water infused with artificial color in order to create the vivid hues of his arrangements, like the bright blue of the moth orchid (*Phalaenopsis*) pictured above.

Inspired by the floral traditions of funerals, the arrangement of chrysanthemums, orchids and anthuriums pictured right was "meant to create a feeling of death," says Værnes Iversen. In keeping with the theme, he left the arrangement on show as the flowers began to die.

"You don't need an education in this line of work.
I believe that people should just jump into it and give it a go."

Værnes Iversen grows his own plants in
a small room within his studio. He uses a
pink light, pictured above right, to give
the plants only the wavelengths of light
they need the most—blue and red—
during photosynthesis.

PHILIP DIXON

The home of photographer Philip Dixon is a contradiction: Intended to be his personal refuge from the outside world, his garden and indoor-outdoor way of life have drawn people from far and wide hoping to catch a glimpse.

At Fifth and Westminster avenues in the heart of Venice, California, stands an old, drab brick storefront. Its big windows have been filled in with painted concrete blocks, and unremarkable additions spread it over four urban lots. Beside the front door, a manicured pygmy date palm stands amid three rough limestone boulders—the only hint of the sanctuary just past this dingy wall. Inside lies the home of fashion photographer Philip Dixon, who originally designed the house in 1993 as a private retreat from the fast pace of the city. Its unassuming, windowless perimeter wall protects a quiet space for Dixon to live and work amid his enclosed desert garden.

Through the brown metal door, a hallway opens into the home's main living space, which is built with thick adobe walls and exposed timber ceilings. Deep niches in these walls and throughout the house hold Dixon's collections of ceramics, books, plants and found objects. These become arresting set pieces against the textured walls, bold complements for portrait photography. Dixon devised the house as a secluded living space encompassing a photo studio, where for years he cultivated his bold style of intensely personal portraiture. Over the years, he has opened the house to other photographers, who have used its walls, pools and gardens as powerful complements to their own work.

From the living space, huge pocket doors slide into the walls to open the room to the garden outside (there are almost no windows in the house). One of these doorways frames a serene view toward a narrow rectangular pool and a raised garden bathed in sunlight. Succulents form a bold, sculptural tableau against the high outer wall. Smooth-barked aloe and Canary Island dragon trees reach over an understory of blue agaves and yellow candelabra cacti. Their shadows scrawl across the smooth concrete floor—an insinuation of heat from the cool depth of the interior. The strong shapes in this garden echo Dixon's artistic treatment of human forms in the light, and the inhospitable image of the desert becomes instead an invitation.

Thick travertine stepping-stones cross the clear blue-green pool toward Dixon's kitchen and dining room. The same thick monochrome adobe reflects and softens the light. The dining area opens to a shaded porch, where a fireplace promises flickering warmth on cool nights, and a built-in couch with thick, rumpled cushions offers a place to sprawl in the shade beside a porous wall of Mexican fence post cacti. Here the harsh and the hospitable characteristics of Dixon's retreat come together most vividly.

Cool light shimmers up from the pool. Faraway sounds of the city barely drift in over the high wall. A short stair climbs to another raised garden and a hidden dining grotto. A path winds among the prickly pears and Thompson's yucca trees, aloes, agaves, fan palms and Canary Island spurge to a large copper door that keeps the outside at bay.

A dipping pool is ensconced in the sort of succulent vegetation that is native to the desert regions from which Dixon took his design inspiration, such as the fan palm and cacti varieties pictured opposite.

Elements of Dixon's home are constructed in adobe—an ancient building material consisting of dried mud bricks. Mud bricks are fireproof, biodegradable and non-toxic, making them popular choices for outdoor building in arid climates.

GUILLEM NADAL

The woodland and wildflower meadows that brush up against the windows of Mallorcan artist Guillem Nadal's studio not only inform his artistic output but often end up being physically incorporated into the canvases he creates.

In the work of artist Guillem Nadal, nature takes center stage. "My approach is very organic," he says. "I look at what's around me—the landscapes, textures and natural materials—and integrate these elements into my work." The canvases, with their gray-black flowing lines, are reminiscent of rippling water. Sometimes, the oil paint is laid on so thickly that it becomes claylike, as if taken from the soil of the earth.

Nadal lives and works on the outskirts of Son Servera, a small village on his native island of Mallorca. A mosaic of wild olive groves, terraced orchards and wildflower-filled meadows surrounds the artist's white adobe house and studio, creating a typically Mediterranean tableau punctuated by the occasional stone wall or rocky outcrop. "I believe that where you live, where you wake up and eat every day, naturally influences what you do," explains Nadal. "My work takes place inside the studio, but the immediate surroundings are constantly with me." Nadal enjoys the untouched quality of the Mallorcan scenery that envelops his terrace, where plants are left to reign freely and the natural light is abundant. His work reproduces the rough textures and earthy color palette of the landscape on his doorstep and explores how they shift with different light or weather conditions.

Nadal's body of work is informed by the five elements of nature—earth, water, fire, air and space—and incorporates raw materials found in his natural surroundings, such as wood, plaster and metal. La Mirada del Foc, a series of artworks dating from the 1990s, features real and drawn-on twigs on white canvases that Nadal burnt in places. The relationship between fire and earth has been a career-long obsession, as has the vulnerability of mankind when faced with the elements. "I want to explore the dependence of man on nature, despite its destructive forces," he explains. "For many people, painting is just about technique, but for me it's important that my work stirs something inside the viewer. Having no message leaves a painting soulless."

Nadal has held solo exhibitions at contemporary art museums and commercial galleries across Europe. His paintings and installations also form part of renowned private and public collections, such as the Würth collection and the Fundació Pilar i Joan Miró. Though the work has found a global following, it is very much rooted in the everyday sights of Nadal's home. "I like the fleeting quality of my natural environment," says Nadal of the ever-changing spectacle unfolding outside his studio as the days and seasons pass. "My art would surely be very different if I lived in Berlin or Paris."

Nadal planted wild olive trees (*Olea oleaster*) and European fan palms (*Garballo*) around the terrace "to create a forest and imitate the landscape" into which the garden eventually segues. It's a low-maintenance solution: The trees are able to endure Mallorca's hot summers and drier winters and, he says, only need to be fertilized once a year.

Artwork can be found at various stages of
completion at Nadal's studio. "They are
a little diary of my thoughts," he says of
his creations. He takes larger sculptures,
like the bronze work titled *Illes del Sol*
pictured above, to the rooftop to dry.

"My work takes place inside the studio, but the immediate surroundings are constantly with me."

The landscape's native flora creeps over the drystone wall boundaries of Nadal's garden (opposite), including prickly pear bushes (*Opuntia*) and the ever-present wild olive trees, some of which are centuries old.

TAICHI SAITO

To create his Tokyo garden, landscape designer Taichi Saito introduced just enough tropical plants to ensure year-round greenery. The rest, he left to one simple creative trick: the principle of *shakkei*—borrowing scenery from his neighbors.

A wooden door opens into the private home of Taichi Saito in Tokyo's Todoroki Park. Trailing plants from a garden terrace above wave quietly in the breeze, and ferns spread in the shade. Water trickles in the stillness. The fine subtlety of this composition is easy to miss. It was designed "to express dynamic *wabi-sabi*," says Saito, who owns landscaping studio Atelier Daishizen and conceived his own garden. He explains that he was seeking to capture a sense of fleeting, imperfect beauty.

Inside, a low mezzanine overlooks the living space toward more ferns, palmettos and bird-of-paradise plants in the moist shade beyond. Saito selected plant varieties from around the world to ensure perpetually green foliage and "flowers and fruits in all four seasons." Despite close neighbors and the home's huge panes of glass, there's a serene sense of privacy.

The home was designed by architect Tsuyoshi Tane and completed in 2018. Saito explains that the project assumed an "aura of inevitability," because there was only so much they could do to the site. Tane considered primitive architectural responses to climate, and then observed the natural order of the compact site, with its low, shaded north side and drier, more exposed southern aspect. He settled on creating

two slightly irregular eight-sided parts—two houses, really, one for dry conditions above another for wet conditions, reflecting the dual climate of Todoroki Valley.

The lower, earthbound living space expands toward a north-facing garden. In the sun-dappled living area, mid-century furniture accentuates the earthy warmth of the space. The room sits partially belowground, and its low horizon extends to the roots and trunks in the garden while the rough-cast walls turn in to form a space for dining. Tane designed these earthbound walls to keep the space cool in the summer as warm breezes drift through open windows. In the winter, warmth radiates from the walls and heated stone floor to the sleeping rooms above.

The open stairway climbs toward the bright main bedroom in which niches with potted philodendrons and large windows present the foliage of neighboring trees—borrowed, Saito notes, in the tradition of *shakkei* (incorporating background landscape). A door swings out to a terrace over the entry court. It is evident why Saito says his house "gets along beautifully with the wind." The dry air blows, bringing with it the soft sounds of stirring leaves and, quietly, almost forgotten, the noises of the city.

Saito grows potted ginger as a house-plant, pictured above. Ginger is a reed-like plant that can grow up to three feet (1 m) tall and produce yellow flowers.

LAS POZAS

The concrete garden of *Las Pozas* had been all but swallowed by the unruly Mexican jungle at the point at which it was rediscovered by conservationists. They decided that the subtropical invasion only added to its surrealist spirit.

At the Jardín Escultórico Edward James in Xilitla, Mexico, spindly concrete structures unfurl over seventy-four acres (30 ha), entwined by dense, green foliage. When walking through the thick undergrowth with the forest canopy high above and natural waterfalls as a backdrop, or else scaling the tiered vine-wrapped platforms, the garden feels less like reality and more like a magical, M. C. Escher sketch come to life.

Known as Las Pozas ("The Pools"), the garden was imagined and constructed by surrealist artist-poet Edward James over three decades. It was back in 1949 that James and his team of local builders first broke ground. They fashioned sweeping turrets, skinny walkways, wall-less rooms and tiered lookouts out of concrete. The dreamlike buildings, with names like The House on Three Floors Which Will in Fact Have Five or Four or Six, were as much about the natural world as human creativity and designed to become entangled with plant life over time. James brought in dozens of varieties of plants from other regions; at one point, there were thousands of orchids on the premises.

Fundación Pedro y Elena Hernández, a foundation conserving Mexico's natural protected areas, acquired the grounds in 2007 in order to steward the land, preserve the garden's natural and cultural heritage and open the place to the public. When the foundation took the reins of Las Pozas, it had been mostly untended since James's death in 1984. The climbing plants, swelling roots and humidity had been slowly disintegrating the concrete for twenty years. Epiphytes had swarmed the buildings, and the whole site was close to being consumed by the surrounding subtropical jungle. The foundation did a full diagnostic assessment of the grounds and the structures before proceeding with any interventions.

"The extreme ravages of nature have caused some damage, oxidation and crumbling, but it's not as serious as it looks," says Frida Mateos, coordinator of conservation for the foundation. "From a purely engineering point of view, James's structures are capricious," she says. "Nevertheless, the local workforce had an empirical and intuitive body of knowledge and they built the structures in an extraordinary way."

The foundation developed the infrastructure that allowed the grounds to reopen to the public. "We call the space a garden to refer to James's interest in gardening, and so visitors understand they are visiting an outdoor place where plants carry a major role," says Ximena Escalera, the program manager at Las Pozas. "On the other hand, a garden is a place where the intervention of man can be seen clearly," she adds. "Here, the jungle plays to James's hand—it's exuberant and perpetually growing."

James eventually created over thirty architectural follies at Las Pozas, including the eye-shaped tub (pictured opposite), which he sunk into a patch of ferns, banana plants and heliconia species, and the site's circular main entrance—known as the Queen's Ring, pictured above.

The Stairway to the Sky, pictured above, is perhaps the best example of what James himself called his "pure megalomania." It has presented conservationists with a conundrum over how to maintain its charming decayed grandeur while ensuring its concrete remains strong enough for visitors to climb.

"A garden is a place where the intervention of man can be seen clearly."

The Hands, a sculpture of two right hands seen from both sides, hints at James's fascination with surrealism. Before he arrived in Mexico, he had sponsored Salvador Dalí for the whole of 1938 and appeared in two paintings by René Magritte.

MARISA

COMPETELLO

In 2015, Marisa Competello followed a "random and decisive" impulse to reinvent herself as a floral designer. Encouraged by her background in dance and fashion styling to create bold compositions, she found that her business blossomed.

Early mornings are an unavoidable part of the job for New York–based floral designer Marisa Competello. By seven a.m., she's already powering through Manhattan's Flower District to gather stems for her striking, minimalist arrangements. "It's this block filled with bounty. All the stores are overflowing with plants and flowers out onto the street," she says. "Typically, every day is a mission."

Competello then returns to her Chinatown base to start crafting the sculptural floral creations on which she's built the reputation of her studio, Metaflora. Forget any preconceptions of fussy, overblown bouquets: Competello's creations often comprise only one species placed into a highly stylized form. "I see them as more masculine," she says of her designs. For one recent creation, she arranged a fan-shaped collection of spiky Bismarck palm fronds, softened only by twisting pieces of *Cytisus* that gently curl like strands of flyaway hair.

Competello used to be a dancer, then spent nearly a decade assisting fashion stylists. She says that this unconventional background provided foundational training for floristry. "Dance comes into play because I've been studying lines and shapes my whole life," she explains. "You can see that in my work—the designs are more sculptural at times, and you can see movement. And my styling background laid the groundwork for me to put things together based on color, texture, silhouette."

Her decision to move from fashion styling to floral design didn't come as a total surprise: She'd sold hand-wrapped bouquets to neighbors as a child, and later dabbled in flower arrangements for a friend's restaurant. Still, she describes the actual moment of transition as "quite random and decisive."

Metaflora's first client, a Lower East Side café called Dimes frequented by the neighborhood's hippest residents, proved an invaluable launch pad. "It became this amazing community . . . a lot of people saw my arrangements there," she recalls. "I got to experiment and get weird because there were no boundaries." After that first big break, her business grew, and her résumé now includes a 2019 faux flower collection for home decor retailer West Elm and events for Nike and design studio Apparatus.

These days, Competello often finds herself proffering counsel to amateur hobbyists enamored by her work. Her advice is always the same: Visit a flower market and see what catches your eye. Although she has her own roster of favorites ("I like these velvety, lobster-looking flowers called heliconias"), she encourages people to follow their own creative inclinations. "Put together whatever you're drawn to. There are no rules," she says. "That's how I've created my aesthetic. Draw outside the lines a bit, and you'll end up with something interesting."

Above, an arrangement of kangaroo paw (*Anigozanthos flavidus*) and green anthurium decorates a podium in the Metaflora studio. Kangaroo paw is native to Australia. In the wild, its sturdy stem is a natural perch for birds. Opposite, Competello holds a single red parrot tulip (*Tulipa* × *gesneriana*).

The ruffled head of a parrot tulip (*Tulipa × gesneriana*), pictured below, sits heavy on its stem—they can often grow five inches (12 cm) across. Opposite, a papyrus sedge (*Cyperus papyrus*) softens a corner of the Metaflora studio.

Competello often selects just one species to create her highly stylized sculptural arrangements, like the coconut cluster centerpiece pictured above.

SOURABH GUPTA

"Fake it to make it" is an established career trajectory in New York City. It's certainly how artist Sourabh Gupta makes his living—except what he fakes are botanically accurate flowers, and how he makes them is nothing short of an art form.

Sourabh Gupta is a designer, architect and artist based out of a studio in Harlem, perhaps best known for the paper flowers he makes there. His path into this unusual practice was as organic and wild as the flowers he replicates, but it all started with a garden. As a boy in Kashmir, Gupta created a verdant rooftop garden, replete with thirty-three bonsai trees and hundreds of plants (which his mother now tends while he's in New York). Because there was nowhere to buy pots for the plants nearby, he made them himself. "Everything that I wanted to create, to decorate, I had to make myself," he says. "That's where the concept of making something started in my life." He accepted plants and flowers from friends' gardens and grafted them onto plants in his own, which, he says, "gave me time to grow and surround myself with nature" and "to study and understand it." Inspired, he made paper flowers to decorate the church at his Catholic school.

Gupta moved away from home to study architecture in India before moving to New York to study at Parsons School of Design. There, he met designer Stephen Earle, who one day took him to the Metropolitan Museum of Art to see an exhibition called *Heavenly Bodies: Fashion and the Catholic Imagination*. He said the textiles reminded him of the paper flowers he made as a child. After the exhibition, Earle gifted Gupta some paper, and Gupta started making flowers again—and it filled his life. "There's so much diversity in flowers, so much expression, that you hunger for more—to find the next color, shape, smell. There's something new and fresh about every single flower." He gave his constructions away to his friends, who loved them because they could "survive" in the lightless rooms of their New York apartments. "Nature was there with them," Gupta says.

His interest in re-creating botanically accurate flowers (as opposed to purely decorative ones) intensified after a particularly inspirational trip to the gardens of the Hamptons. "It was too much for me to understand—in terms of color, anatomy, structure. And, of course, my architecture side always comes in. When I see something, I see it in parts. When I see that many components in flowers—that many layers—it's even more enticing for me to know how it is put together." After the garden tour, he spent his whole Christmas holiday constructing a single thistle plant.

Now, his architectural and botanical interests have dovetailed. He has dozens of books on botany and spends a lot of time thinking about structure as beauty. "There's so much in the gesture, how pieces are attached, the scale. There is something magnificent about a whole plant. It almost feels like poetry." (Right now, he says, he's most interested in the wild oak leaf hydrangea.) And, like any good designer or artist, he's interrogated his own fascination, leading him to more existential questions that verge on Shakespearean. *What makes a flower a flower*, he now wonders—*if it had any other smell, shape, or color, would it still be a flower?*

The path forward for Gupta, it seems, will be as whimsical and inspiring as his childhood garden high above the Kashmiri streets.

Before Gupta creates a new flower, he makes careful studies of its petals, palette and the materials that might best suit it. "Each drawer becomes a laboratory of that particular flower," he says. The orange hawkweed (*Hieracium aurantiacum*), pictured above and considered an invasive weed in several US states, was drawn for a client in advance of making the model.

Gupta recently exhibited twelve cre-
ations inspired by the Adirondacks' wild-
flowers at the Bolton Historical Museum.
Pictured below, he creates the stem of a
Carolina rose—a common shrub in North
American pastures.

The lotus plant, pictured opposite, is designed to mimic an ikebana arrangement. Gupta says he was inspired by flowers he saw when he was invited to visit the private garden and pond of Jay Johnson (brother of interior designer Jed Johnson). "I was absolutely stunned as I stepped into that garden," he recalls. He is fond of the geranium, pictured below, for the way the plant grows as it seeks light. "It produces amazing poetic gestures," he says.

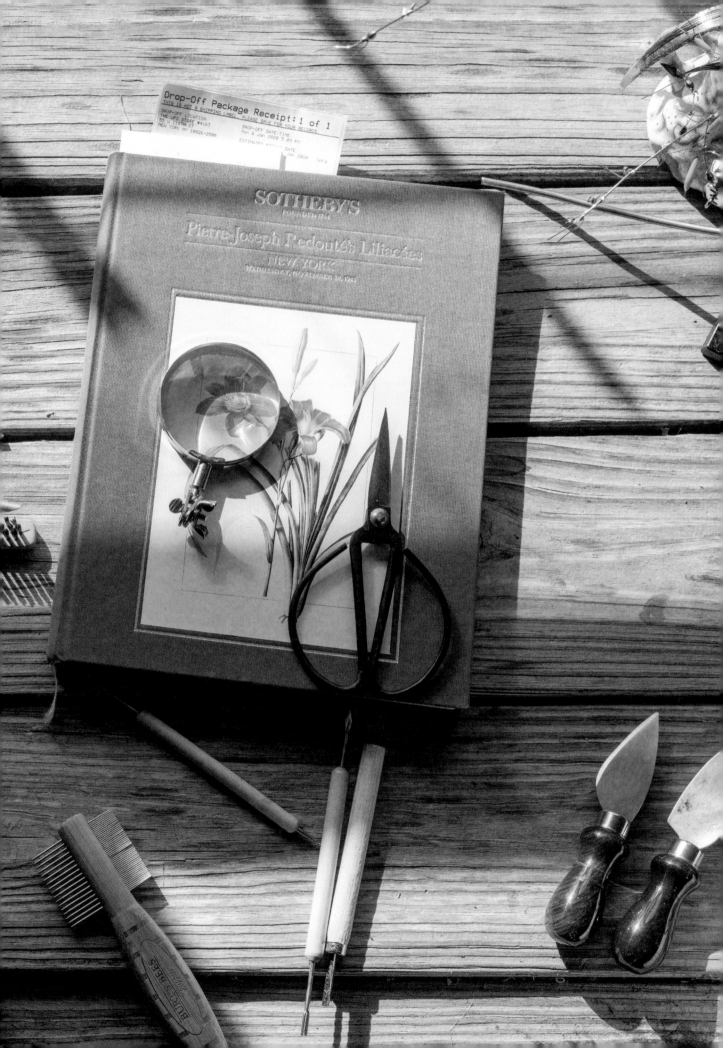

SOTHEBY'S
FOUNDED 1744

Pierre-Joseph Redouté's Liliacées

NEW YORK
WEDNESDAY, NOVEMBER 19, 1985

"There is something magnificent about a whole plant. It almost feels like poetry."

Although the art of making paper flowers is centuries old, Gupta does not follow its traditions. He uses paper towels to make stems and toilet paper to create stigma and colors each bloom by hand with Chefmaster Liqua-Gel food coloring.

CÉCILE DALADIER

Paris-born ceramicist Cécile Daladier makes strange, sculptural vessels designed to hold delicate wildflowers. In work and in life, she's inspired by the bounty of the countryside that surrounds her remote farmhouse in *Drôme*.

A native Parisian, artist Cécile Daladier grew up in a town house overlooking Notre-Dame cathedral and the Seine. There, her father grew a leafy courtyard garden and cageless birds roamed around her bedroom. She spent weekends and holidays at her grandparents' country house, enjoying long walks in the surrounding forest and helping her grandmother tend to her flower-filled garden. Being close to nature is a cherished memory from her upbringing. "Nothing specific sparked my love for gardening, it was always there," she says.

Today, Cécile lives with her husband Nicolas Soulier (an architect), a dog and two cats in a stone farmhouse perched at an altitude of a half mile (800 m) above sea level in the southern French region of Rhône, where grassy pastures meet rocky mountaintops. The couple first acquired the property in 1985, and spent years refurbishing it; the transition between city living and isolated idyll was gradual.

Cécile explains how, at first, making use of the property's seventy-four acres (30 ha) of land required a certain level of creativity. Formed by steep slopes rich with wild vegetation and limestone soil, land suitable for gardening was limited and only specific species were fit for planting. "It's not a garden in the traditional sense," she says. "It's more of a wide-open space." Although she takes care of the trees and the weeds that grow on the steep hills, Cécile has created a strip of cultivable land along the walls of her house where she can let her passion for gardening loose.

Here, she mixes species from the surrounding fields with those she finds at local nurseries. "I like to go from wild to cultivated," she explains. She recalls once planting a blue columbine—a plant with small, colorful flowers that grows in the surrounding woodlands—next to a black one, and how, a few days later, a pink one somehow grew among them: "All naturally, like magic."

Cécile has understood over time how the key to good gardening lies in acknowledging and appreciating your surroundings and in letting nature do its work. "My philosophy is to start with what you have and what's already there before you, and build on it," she says. "Before creating a garden, you have to live in it for a few months—maybe even a year—to observe what happens, what grows and what disappears."

She admits how, at first, she would tear out wild plants and flowers that, despite being unique and native to the region, she perceived as invasive and unsuited to a well-kept garden. "We made many mistakes in thinking that we knew better," she laments. "You simply have to wait. Nature is a present. It's beautiful left alone."

Today, life in the couple's farmhouse follows no regular routine. Cécile works in her spacious atelier, or else passes the time by arranging flowers (she recently put together a few branches of the purple-flowering wisteria that grows outside her front door), playing the piano or reading (mostly Marcel Proust). One thread that binds her days together, however, is her relationship with nature. "I work with plants every day. I do gardening, help the flowers that are blooming, and water them if necessary," she says. "But sometimes, I simply stop and observe the beautiful nature that surrounds me. It gives me ideas for my work and space for my creativity to breathe."

Above, Daladier's home studio features
many potted plants; her artistic practice
is, she says, informed by the elements and
processes of the natural world. Here, she
creates abstract paintings, environmental
art, installations, objects and gardens.

The garden changes dramatically in summer. The May-flowering wisteria that Daladier planted on either side of the front door blooms with scented purple flowers that creep over the stone walls of the farmhouse. A lighter, second crop blossoms again in August.

Daladier's small vases, which she calls *pique-fleurs*, are intended to give individual flowers more attention. When firing ceramics, she sports the old French firefighter's helmet pictured opposite.

Surrounded by mountains, with views stretching to France's Rhône Valley below, the farmhouse was the ideal place for the Daladiers to escape the city. The Drôme region is historically known for its pottery (particularly the nearby towns of Cliousclat and Dieulefit).

Six Tips

A bouquet of flowers breathes life and brings color into any living space. It is a *tableau vivant* that evolves over the arc of its brief life, from budding promise to droopy decay. As artful arrangements can often be expensive to maintain, the tips over the following pages offer ideas on how to make the most of what you already have available, whether by prolonging the life span of cut flowers with careful tricks, by foraging your own arrangement in the wild or by drying and pressing a bouquet for posterity.

How to Create with Flowers

Words by Amy Merrick

Author of *On Flowers: Lessons from an Accidental Florist*

1. HOW TO EAT FLOWERS

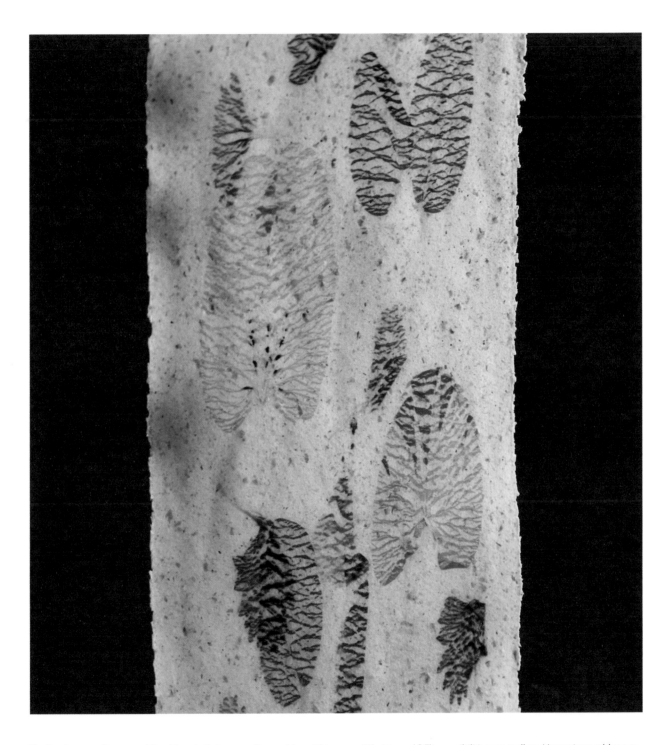

The line between flowers and food is an indistinct one: The most beautiful things always seem nearly edible. The impulse is understandable, as we often eat plants and their fruits, roots and seeds. So why not revive the practice of eating flowers? When thoughtfully selected and sourced, they add beauty and flavor, not to mention nutrition, to your plate.

Not all flowers are edible, though; it's safest to use varieties that have stood the culinary test of time, and to source only blooms that you know to be pesticide-free. Make sure to wash and dry any petals before preparation, too, to remove soil and loose pollen.

What to cook? The possibilities are endless. Nasturtiums add a peppery bite to salads, and can be pressed into pasta for flair. Elderflower can be distilled into a cordial or battered and fried into golden, lacey blooms dusted with powdered sugar. The petals of roses can be crystalized or distilled into rose water. Squash blossoms or daylilies can be carefully stuffed with ricotta cheese and gently fried: Their flavor is mild and summery. Flowering herbs such as chive blossoms can add an allium note to salads and infused vinegars. Borage, pansies, lilac, violets and geranium petals are all mild enough flavors to bedeck any cake.

2. HOW TO DRY AND PRESS FLOWERS

Drying Flowers

Pressing Flowers

Rather than simply letting flowers dry or wilt in the vase, intentionally dry them for a lasting impact. Flowers should be dried while still very fresh, not after they've given their all in a bouquet, to keep their shape. Make a bunch and tie the stems with twine, then hang them upside down in a dark, dry, preferably warm place with good airflow. An attic or a garden shed is ideal. Leave bunches undisturbed for two to three weeks until they are entirely dry. Most flowers retain their structure and color, but others will fade. Prime picks for drying are strawflowers, celosia, hydrangea, lavender, lunaria and yarrow.

There are few things more delightful than happening upon a pressed flower in an old book: They are nostalgic reminders of moments in time. Find a heavy volume like a phone book to use as a press, as its pages may wrinkle with moisture. Use delicate flowers with minimal structure, such a pansies, violets and primroses to press, as opposed to thicker blooms like roses or tulips, which would require a large mechanical press to flatten. Line pages with newsprint and lay flowers flat. Weigh down with more books and put in a warm, dry place. Check every few days until flowers are paper thin and dry.

3. HOW TO ARRANGE IN SPRING & SUMMER

Spring

Summer

Spring's inspiration starts small: Minuscule flowers shiver through the cold while buds on branches swell into the tiniest of green leaves. The charms of these early blooms, like snowdrops or muscari, shine when styled demurely in glass, seemingly full of rain. Soon, though, that drip turns into a torrent and blooms burst forth, rioting up from the ground in a shower of petals. Bulbs bloom in drifts, and gardens burst with roses: Late spring is irrepressibly vibrant. Make like a Dutch master and celebrate the spring bounty with a painterly array of colorful blooms.

The flowers of summer are deeply unfussy: a meadow of wildflowers caught in the breeze, sky-high sunflowers cut from your local farm, a jam jar of daisies haphazardly strewn on the kitchen table. A loose, airy bouquet and a rustic, repurposed vase refresh spirits when the sun beats down. Summer markets tantalize with buckets of favorites like zinnias and dahlias, but the heaving tables of fruit and vegetables can also yield elements worthy of a floral display. In summer tomatoes on the vine, fennel flowers, or herbs like rosemary, purple basil and chamomile can beautifully go from plate to vase.

Fall

Winter

Fall arrives as a relief to most, and with it comes texture and drama in the natural world. Rusty rose hips of every size and scale tangle in thickets, milkweed seedpods erupt into billowing downy clouds, blossoms turn to berries, trees hang heavy with fruit, and grains ripen into wavering bunches: deep purple grassy millet or plumes of copper amaranth. Colors burn bright then fade, and flowers like asters, chrysanthemums and antique hydrangea linger until the first kiss of frost. Woven baskets and wooden vessels feel especially apt for the season.

At first glance, winter might seem barren. Nature, though, tells a different story. Only in winter are the frenzied distractions smoothed over to reveal the structure and soul of a landscape. Fragrant evergreens, twisted by the wind, materialize alongside sculptural branches with birds' nests or clumps of lichen and moss. Paperwhite bulbs, amaryllis, hellebores and cyclamen bring color home, while bowls of citrus speak of warmer places. Velvety buds on tree branches can be forced into early flowering when brought inside. Winter blooms with elemental simplicity for those who know where to look.

Flowers die: a sad but true fact of life for all those who love them. Ephemerality is also their greatest gift, a gentle and graceful reminder to appreciate moments as they come. Keeping cut flowers fresh for as long as possible starts before any arranging takes place. Vases, buckets and clippers need to be kept scrupulously clean to avoid introducing any bacteria into the mix, which would shorten the life of blooms. Stems should be clipped sharply at a 45-degree angle to aid in water absorption, and all foliage below the waterline should be removed to avoid rotting and the dreaded stench of a bouquet left alone to fester.

Cut flowers have no special love for sunny spots; a hot, sunny corner at home will actually speed along wilting. Opt for cool, shady spots and use cool water to refill vases as needed. There is nothing so stingy as giving flowers only a few inches of water to drink; they will be happier if their vase is full to the top. When flowers start to wilt naturally, certain delicate varieties may go first and it's often easiest to remove them and remake the bouquet into something smaller, changing the water completely and giving all the stems a fresh cut. Some varieties will last just a few days and others for weeks, but the joy they give is without measure.

A flower arrangement doesn't need to be flowery to be beautiful; in fact, it doesn't even need to contain flowers at all. Opening up your vases to a world beyond blooms inspires creativity, sustainability and awe in the natural world around you. Arrangements that feature foraged materials inherently reflect the season and environment.

Sculptural bare branches have a structural allure that speaks volumes through negative space. Pillowy, lush mosses can charm in a low, shallow dish, creating a microcosm of the forest floor on your table. Wisps of feathery grasses that on the roadside seem banal can create a billowing meadow indoors when showcased each as specimens in separate bottles of all different sizes. Foliage with unusual colors like metallic silver rex begonias or fiery tie-dyed coleus hold their own in a vase, while oversized palm fronds or banana leaves add a tropical statement when brought inside.

Be respectful when sourcing these nonfloral stunners, though: Only forage materials from your own garden or that of a friend, never from public places without express permission. When cutting, only take what you need from plants that have enough to spare, and afterward, it should look as though you had never been there at all.

Vases

Change a vase's water every two to three days. The stainless steel pipe vase pictured right was designed by Tableau and Bloc Studios to be easily detached from its raw stone pedestal for quick freshen-ups.

Simple terra-cotta pots "breathe," which makes them ideal for novices who may overwater plants. Glazed terra-cotta, like Skagerak's amphora-inspired pot, are less porous but as a consequence retain water for longer.

Put fresh cuttings or new seedlings in small vessels, like Svenskt Tenn's Acorn vase, with fresh water and watch them begin to sprout.

Clear glass vases like Warm Nordic's Arctic vase (originally designed in the 1960s) are timeless, and make perennially good pairings for classic arrangements.

The size of a vase dictates the scale or type of arrangement it can hold. Try to find vases with different-sized openings—like Kristina Dam Studio's Dome series—if you wish to experiment with single stems or branches.

Combine vases of varying sizes and colors, like Iittala's geometric Ruutu vase series, to complement and juxtapose different floral compositions.

PART THREE

Community

It takes many buds to make a flower bed; it takes many hands
to make a garden. Meet the global communities blossoming
alongside the nature they tend.

BABYLONSTOREN

In the shadow of South Africa's purple Simonsberg mountain, *Babylonstoren* farm
resort invites guests and volunteers to harvest its eight acres (3 ha) of organic gardens and
attend community workshops on everything from flower pressing to plant foraging.

Nature is symbiotic by design at Babylonstoren, the eight-acre (3 ha) farm, garden and hospitality complex in South Africa's cartoonishly beautiful wine country. To wit: The carrots and onions are planted side by side because each wards off the pests attracted to its neighbor. Most paths for strolling are gravel, but some are charmingly paved with discarded peach pits. Every day, ducks are ushered into the fruit orchards to gobble up snails hungry for the trees' tender leaves. The plants are spritzed with an organic pesticide of wild garlic, chili, basil, and marigold, grown on the premises and personally prepared by veteran Babylonstoren gardener Gundula Deutschlander.

There have been European-style gardens at the property since 1692, reflecting the Western Cape region's history as a "feeding station," as Deutschlander puts it, for European sailors and soldiers rounding southern Africa en route to Asia. In 2007, owner Karen Roos commissioned French architect Patrice Taravella to create the bones of Babylonstoren. Their collaboration has borne delicious fruit: Today, guests can wander through rows of fragrant flowers, vines, trees, hedges and a prickly pear maze, pause under trellises of hanging squash or lie down in a summer chamomile field designed for horizontal luxuriation.

At Babylonstoren, Deutschlander and her colleagues have pushed to "celebrate what we are today" by bringing native plants off the veld for cultivation in traditionally European walled gardens. Babylonstoren has also expanded the garden team, as well as launched on-site training for people who want to develop their home gardening skills.

Visitors also have access to community workshops, like flower pressing, indigenous plant foraging and fermentation. "It's a balance between wanting to share information and wanting to learn," says Deutschlander. "We look for what each person can contribute and draw that out."

Most of the indigenous plants at Babylonstoren are medicinal, says Deutschlander, with some tasty exceptions. "We just had a harvest today of num-num berries, botanical name *Carissa*," she says. "It's a lovely fruit, high in vitamin C and makes a wonderful preserve—a bit like a cranberry."

Deutschlander, for her part, says she gets as much energy from gardening as she gives to it. "Gardening has been incredibly generous. It makes me feel so alive. Like today, I was walking past a volunteer picking basil and I stopped to admire the bees frolicking in the pollen. Every day I get something back and feel stronger. It makes me feel like I have my little bit of power to make a difference."

Gundula Deutschlander has been tending to Babylonstoren's garden since its inception. She is also guardian of the site's Healing Garden, which includes plants used in African healing traditions, such as the sausage tree, African flame tree and stinkwoods.

"Gardening has been incredibly generous. It makes me feel so alive."

Above left, cut dahlias are kept in water ready to make bouquets for any visitors celebrating anniversaries at Babylonstoren. Above right, sales host Racheal Samyels creates a pressed arrangement out of African wormwood (*Artemisia afra*) and moringa.

In between rows of fennel and basil, master vegetable grower Darryl Combrinck embraces Thandi, the scarecrow. Pictured opposite, head gardener Liesl van der Walt demonstrates the simple but effective tool used to harvest fresh prickly pears (*Opuntia*)—a tin can nailed to a long stick. The tin is placed over the pear and twisted so that it falls from the tree. The picked pears are washed, peeled and served chilled.

Opposite, master botanist Ernst van Jaarsveld holds a pair of twisted kudu horns aloft. Above, the master of fruit trees, Wendell Snyders, takes a rest with a prized pumpkin. The garden at Babylonstoren contains more than 300 species—all of them edible or medicinal.

KRISTIAN SKAARUP

&

LIVIA HAALAND

Every Wednesday, members of ØsterGro climb to the rooftop of a Copenhagen office block to harvest produce and discuss goings-on at their community farm. For founders Kristian Skaarup and Livia Haaland it's not just a meeting place, it's a model.

In Copenhagen, a surprising addition to the historic cityscape sits on the rooftop of an old car auction house: a 6,500-square-foot (600 sq. m) farm. More than 100 varieties of plants are grown here, from herbs, salads and fruits to root vegetables and garlic bulbs. Living alongside these rows of carefully tended plants are nine chickens, two rabbits and however many bees it takes to fill three hives.

"Growing food in the city appeals to me," says Kristian Skaarup, who cofounded the farm with Livia Haaland in 2014. ØsterGro is a first for Denmark, inspired by the collective spirit of pioneering, community-run rooftop gardens in Brooklyn and Seattle. Members pay a biannual fee of approximately $270 and, in return, receive a weekly supply of freshly grown vegetables. "You build a relationship with local people," Skaarup says of the rapport that develops when members show up each Wednesday afternoon to collect their bounty. From April until December, volunteers join every week to harvest produce from the garden, package the vegetables for collection and manage the compost.

Up to five evenings per week, the cozy twenty-five-seat restaurant Gro Spiseri, housed in ØsterGro's greenhouse, invites diners for a seasonal menu sourced from the rooftop and other local farms. And each day, at two p.m., the team of volunteers sits down together and shares lunch. "Hopefully, they also gain a network," Skaarup adds. "There are a lot of international residents coming here, who may not know many people."

A landscape architect by training, Skaarup has always been fascinated by the green enclaves carved out in urban environments. Copenhagen, with its abundance of parks, offers open space for its citizens, but they're carefully regulated: "You can't really do anything in them," he says. In contrast, Skaarup and Haaland want to create a participatory space that encourages the city's residents to learn about food production through hands-on initiatives. "For the volunteers, it gives them a day of doing something else beyond being in the city," observes Skaarup. "It's something different—messing with soil, seeing something grow and then harvesting it. It has a healing effect."

Beyond the curative powers of community and cultivation, Skaarup hopes that ØsterGro can serve as an example of regenerative food production. He is realistic about the globalization of the agricultural industry and the limited impact that his small rooftop farm offers. "We can't grow all our food on roofs in Copenhagen, it's not possible," he says. Instead, he hopes to reform current views on food production by showcasing a more thoughtful attitude toward consumption. The principles illustrated at ØsterGro—buy local, seek out organic produce, reduce your food waste—are actions Skaarup hopes the city's residents will be encouraged to implement. "This is more of an inspiration," he says. "It's not about the quantity; it's about the quality."

Pictured opposite, volunteers sit down to a *fællesspisning*, or communal lunch, by ØsterGro's sprouting beds after a morning's work. Below, small organic bouquets of peonies, yarrow (*Achilleamillefolium*), mallow (*Lavatera*), masterworts (*Astrantia*) and cosmos flowers decorate the greenhouse.

All produce cultivated at ØsterGro's rooftop farm is organic and either sold to community members or served family-style in the greenhouse to guests of Gro Spiseri—the in-house restaurant.

Below, Skaarup and a volunteer tend to the beehive. Opposite, Haaland feeds the chickens in the henhouse with leftover produce from the kitchen and fields.

LUCIANO GIUBBILEI

When gardener Luciano Giubbilei discovered that the traditional home of a Mallorcan potter was on the market, the next step seemed obvious: Turn the house into a creative retreat and invite people to use it as a base to commune with nature.

Award-winning gardener Luciano Giubbilei was raised by his grandmother in a second-floor apartment in the center of Siena, a medieval city in the Italian region of Tuscany. "I wasn't interested in gardens as a small boy because I had none around me," he remembers.

After leaving home, Giubbilei cultivated a vegetable garden. The intention had been simply to enjoy fresh ingredients in the meals that he prepared for loved ones, but he soon realized how much he savored the rituals around growing plants. After apprenticing at Villa Gamberaia—a historic Tuscan manor house that boasts an exquisite terraced garden—Giubbilei relocated to London to study at the Inchbald School of Design. Today, he "makes gardens"—a description that pays little due to the research-intensive process that can see him, in the case of a commission for an estate situated on the Spanish Balearic island of Formentera, invest five years in a project.

"It needs to speak the language of the place," he says of his approach, which dictates that he deeply understands the character and conditions of the landscape before the garden's design is mapped out. "But it also requires using plans that work on every level, aesthetic and practical. It takes some time to refine that."

A regular visitor to Mallorca, Giubbilei was urged to make the acquaintance of a local ceramist one year, who resided in a traditional house with a small walled garden and two terraces. Four months after their meeting, she died, and Giubbilei, again at the urging of friends, bought the property. As renovations progressed, he found himself contemplating how he hoped the home—now named Potter's House—could be used. "I thought it was very important to reestablish the purpose of the place," he explains. "It's about making, and it's about having creative people come here."

His first guest was Maria Kristofersson, a ceramicist from Sweden who stayed for eight weeks. Two more artists are scheduled to visit this year: one a multidisciplinarian whose talents include film, black-and-white drawings, and music; the second, another ceramicist. "I love that this space is their space," says Giubbilei, who purchases part of each resident's output during their stay so that a visual record can live on at the property. "I like that the place is filled with their work and their energy."

Five minutes' drive from the house, down a dirt track, is a field recently leased by Giubbilei. The plot of land, which will be framed by drystone walling and filled with native and exotic plants, will offer guests another area in which he hopes kinship and creativity will flourish.

"When people come to Potter's House, we can share a beautiful moment with them—not just in the house but also in the landscape," says Giubbilei. "We can invite people, eat, and have a moment of communion."

Pictured above, an old pomegranate tree hangs over a garden planted with Mexican feather grass (*Stipa tenuissima*), maidenhair vine (*Muehlenbeckia complexa*) and myrtle bushes (*Myrtus microphylla*).

"We can invite people, eat, and have a moment of communion."

Giubbilei has encouraged climbing plants to feel at home on the terrace, pictured opposite. The climber on the building is Boston ivy (*Parthenocissus tricuspidata*), a deciduous vine that regrows its leaves in spring, and the far wall is host to a star jasmine (*Trachelospermum jasminoides*) and climbing fig (*Ficus pulima*).

KAMAL MOUZAWAK

Kamal Mouzawak could never have anticipated how his mission to put homegrown food on the plates of Beirutis would expand: first a market, then farm-to-table restaurants and now a guesthouse set amid mountainous produce gardens.

Kamal Mouzawak, the Beirut-based sophisticate behind some of Lebanon's most celebrated community initiatives, felt ashamed of the traditional *tannour* that enclosed his sandwiches as a child. "Everyone else had this modern Arabic bread," he recalls of the convenience food that had recently come into fashion among his classmates. Mouzawak, raised in a village twelve miles (20 km) from the capital, was given old-fashioned flatbread and meals that revolved around seasonal produce: grapes picked in the summer, parsley plucked straight from the garden and citrus delivered in bulk by his uncle. "It was very much eating from the land," he says. "It was not a concept, but our daily life."

This firsthand knowledge of local traditions has proved instrumental in forging Mouzawak's reputation as an arbiter of Lebanese hospitality. After establishing himself as a food and travel writer, he launched Souk El Tayeb, a weekly farmers' market, in 2004. Producers from around the country were invited to sell their organic wares at stalls in central Beirut, offering an opportunity for direct contact between rural communities and the city dwellers who relied upon them. "People in cities think that food is a commodity that you buy from supermarket shelves," says Mouzawak. "They forget that someone has produced or planted or cooked it."

By 2009, Tawlet—a farm-to-table kitchen located in Beirut's Mar Mikhael neighborhood—was born. Lebanon still struggled with the legacy of deep-seated political and sectarian divisions, but Tawlet created a collaborative space where a rotation of women from across the country were invited to cook their villages' delicacies. Regional outposts cropped up quickly, encouraging curious diners to venture from the coastal city of Saida to the lush agricultural splendor of the Bekaa Valley.

"In such a diverse place, you need to find common ground between all of these different people," says Mouzawak. "The common ground is not the politics and it's not the religion. The one and only expression that travels through time and place is an expression of tradition. It's cuisine."

This vision of national pride lends itself well to Mouzawak's latest venture, Beit, a series of homes that have been converted into guesthouses. At Beit Douma—Mouzawak's debut foray into hospitality—the pleasures of rural life take center stage. First used as the private home of Mouzawak and his partner, fashion designer Rabih Kayrouz, the eighteenth-century villa was repurposed as a six-bedroom B&B in 2015. After sensitively restoring the stone facade, the couple filled its pastel-hued interior with a patchwork of antiques, framed prints, colorful suzani textiles and floral arrangements.

Outside, a vegetable garden supplies parsley for tabbouleh and Swiss chard for soups. Decorative elements were included, too. "We planted a thousand wild iris—big, white and very perfumed flowers that bloom in spring—and bushes of broom that cover all of the surrounding mountains, and tons of roses that are in their best habitat at high altitude," says Kamal, who also points out the local olive and fruit trees that adorn the property.

Guests are encouraged to make themselves comfortable in the *dar*—a central living area common in traditional Levantine homes—or to assist with lunch preparations in the light-filled kitchen. In the garden, a mud-brick kiln is used for baking *manakeesh*, a delicious local doughy flatbread. Surrounded by beautiful nature and filled with nourishing food, Mouzawak hopes the experience "nurtures their souls."

"It was very much eating from the land. It was not a concept, but our daily life."

Kamal Mouzawak designed the vegetable gardens at Beit Douma in the style of a potager—a sort of country garden popular in France where vegetables, fruits, flowers and herbs are intermingled.

Mouzawak chops parsley for a tabbouleh salad in the kitchen. The pine tree on the table was bought from a plant nursery for use as a Christmas tree back in 2015. "It was half as big," recalls Mouzawak.

Mouzawak sits on the terrace above Beit
Douma's potager. The large bouquet pictured opposite features chrysanthemums
picked from the garden, alongside foraged oak branches.

MONAI NAILAH McCULLOUGH

A pied piper of the millennial houseplant movement, "Plant Mom" Monai Nailah McCullough has nurtured communities—first in New York, and now in Amsterdam—where watering schedules and wellness rituals are given equal weight.

Self-proclaimed "Plant Mom" Monai Nailah McCullough first made the transition from green-thumbed hobbyist to professional horticulturist when she still lived in New York City and worked in visual merchandising for fashion brands. As a native New Yorker, she hadn't grown up surrounded by greenery; nonetheless, she felt an intuitive understanding of how to care for plants—most of the time. "When I first started getting into it, I killed a lot of plants," she admits with a laugh. With each failed houseplant, however, came newfound knowledge. "It encouraged me to keep going and [to learn] how not to do that again."

Fast-forward seven years and McCullough has over 150 plants in her brood—"a lot of children to look after," she notes. A career in visual merchandising has now been swapped for her own business, Plant Mom, which launched in 2018 when she relocated to Amsterdam. Twice a month, she leads hands-on workshops that introduce her community to the basics of urban plant care, from repotting to planting them in upcycled materials. One client, Soho House, has commissioned a series of workshops that have gradually progressed in complexity; the most recent taught *kokedama*, a centuries-old Japanese botanical art that places ornamental plants into little balls of moss-covered soil.

Alongside these commitments, McCullough conducts private consultations on the design configuration and maintenance of indoor plants. Her previous professional experience has proved useful: One installation in a private residence saw a quarter of the property's kitchen wall transformed into a lush sheet of greenery, while another project turned an empty space above a staircase into a riotous splash of color, with locally grown plants potted in recycled milk jugs.

"I find a lot more joy and enthusiasm when I work with individuals in their homes, helping people create a sacred space with their plants," McCullough says. She describes the community that has sprung up around Plant Mom as composed of young professionals, most of whom are interested in improving their well-being. Plants, she believes, provide an opportunity for self-development.

"Caring for plants is the best way to learn how to care for yourself," she says. "If you identify why you can't keep your plants alive, you can also identify a lot of things that may be lacking in your own life . . . Once you begin to collect plants, you make better lifestyle choices and more conscious decisions."

McCullough has proved an appealing poster girl for millennial gardening. Her Instagram account often shows her standing amid a jungle of greenery—a hip urbanite with peroxide-blond cropped hair who has carved out a natural existence in a man-made metropolis.

"What I try to help people understand is that living in a city does not mean you are not in nature," she says. "Everything we do is nature; we *are* nature. Having plants is only an extension of that."

McCullough is pictured at the Hortus Botanicus in Amsterdam. Its greenhouses contain rare tropical specimens, but McCullough finds equal joy in the less showy outdoor spaces; here, she is bending down to observe some early crocus blooms—"the first sign of spring."

The large tree philodendron (*Philodendron selloum*) is unusual in that it can become fully epiphytic in the wild. Although its roots begin in the ground, it may eventually migrate up larger trees until it has no sub-terranean roots at all. It is also a popular houseplant.

The large agave plant in the background, right, has shot up a long spearlike flower. Agave only flowers once, after which the plant dies. The golden barrel cacti (*Echinocactus grusonii*) in the foreground are endemic to Mexico but endangered in the wild.

"Caring for plants is the best way to learn how to care for yourself."

EDUARDO

"ROTH"

NEIRA

How do you build a New Age mecca in the middle of a dense rain forest without chopping down trees? For Eduardo "Roth" Neira, founder of Mexico's Azulik empire, the answer was simple: Build around them.

"Don't let nature be an obstacle," says Eduardo "Roth" Neira—the creator and spiritual guide behind Azulik hotel and Azulik Uh May, a multifunctional museum fifteen miles (25 km) outside of Tulum, Mexico. The museum and its gardens, both of which incorporate the surrounding jungle, are exemplary of Roth's somewhat unorthodox approach to architecture and design.

What looks like birds' nests perched in the canopies are actually human-sized and intended for elevated relaxation. There are swirling tunnels of bejuco vine. Trees puncture the buildings, growing through central rooms and out again through the rooftops. The space feels at once natural and futuristic—an *Avatar*-esque landscape with gaga curves and an undulating fusion of cement and wood. Notably, however, Roth claims not a single tree was cut down during construction. "Honor the obstacle," he proffers. "Let it teach you."

Azulik Uh May was built to further Roth's mission of connecting nature with art and architecture. "What we have lost as a civilization is meaning. It may sound a little strange, but I believe that art is a fundamental right of man and is a way to make life meaningful again," he says.

What does the design process for a Roth project look like? There are, perhaps not surprisingly, no formal blueprints. Instead, Roth and a group of designers, builders and craftsmen begin by meditating on what natural flow and form might work best with the existing environment. "First, we ask permission from nature to allow us to build," says Roth. Then, he says, they go with the flow, weaving vines into thatched walls to look like macramé, raising pale cement teepees and forming walkways that snake from the ground to the jungle canopy.

The museum holds exhibitions of contemporary art along with holistic workshops; one exhibition by a Norwegian artist featured a site-specific olfactory experience that isolated and replicated scent molecules from algae. Azulik is an ever-evolving work in progress. In the works are individual residences for invited artists, a restaurant floating among the treetops and a second museum.

Despite its New Agey leanings, this architectural whimsy hints at a model for future "nature first" developments. As Roth likes to remind museum guests: "It's not the beauty of the building" that allows for an exceptional experience, rather "the connection that it facilitates to nature."

The library was christened the Hall of Stones because of the presence of five large stones—each of which represents one of the five elements of nature. Roth describes the room's purpose as "a space of contemplation."

Sfer Ik, the exhibition space at Azulik, is bordered by a lake. As always, the water feature and building have been built around the natural vegetation, which includes banana plants and tropical cecropia trees.

The cone-shaped ceiling of this living room, pictured opposite, creates an interior garden, "reaffirming that the architecture is one that privileges the contact between nature and human creation," says Roth. Pinstriped calathea (*Calathea ornata*) and devil's ivy (*Epipremnum aureum*) curl up along the tree's branches.

PHIL, DIANNE &

KATY HOWES

On a sprawling ranch outside of Santa Fe, the Howes family is working together to rewild the West in the rugged way that the landscape demands, while upholding the traditions of stewardship deeply rooted in America's ranching community.

"It's really been a family thing," Phil Howes says, speaking from his home in the New Mexican wilderness. Phil, formerly a game warden for the State of New Mexico and now a ranch and game manager on private land, sees himself and his family as stewards, working in concert with the complicated and rich ecosystem that surrounds them. "Everything we do," he says, "is about fixing and repairing the land for wildlife."

Sustainability, for Phil, his wife, Dianne (who is a botanist), and their twenty-two-year-old daughter, Katy, means being aware of what the land needs now and providing a safe future for it—ensuring that the migratory songbirds have seeds to eat and the trout clean rivers to swim in. "There are healthy ways to do stuff," he says of farming and ranching, "and not-so-healthy ways." Phil is always thinking about the healthy ways—about not stripping the land and running cattle over it as ranchers have done in the past. He balances raising cattle with the restoration methods he learned while studying for a degree in wildlife biology with a minor in range sciences (from New Mexico State University, where he met Dianne).

The Howes have seen their patient stewardship over the last eight years change the world around them. For instance, they thinned forested areas and planted native grasses to promote the growth of meadows. The increased meadowlands allow for more snowfall to hit the ground, which in turn allows for more snowmelt to seep into the soil or else trickle into streams to cool rivers, making a more habitable environment for the trout that Phil and his daughter love to fish. They plant trees like pinyon pines and shrubs that provide seeds for songbirds and chipmunks, which, since their management of the ranch began, have returned in abundance—along with wild turkeys, deer and other fauna. The Howes are creating habitats that allow healthy ecosystems to function and flourish.

Though Phil is firmly rooted in New Mexico—he knows all the plants and animals surrounding him—the scaffolding of his philosophy could translate anywhere. "We are the single biggest users of the ecosystem we are a part of," he says. "Therefore, we have a responsibility to it. And when I say 'ecosystem,' I mean the world . . . the oceans and the icecaps and the Rocky Mountains and the Sahara Desert. All of it."

When asked what advice he would give to people who, unlike him, haven't grown up in the wilderness and who can't ride a horse to the nearest trout stream, he answers immediately: "I would want people to not think of land stewardship as a job." Being with nature is about awakening the creative or artistic side within you, he explains. It's about seeing what's around you, understanding its impression on you. "And that grows you. When you grow, you get better."

Katy prepares to ride the Galisteo Basin Preserve—a conservation area fourteen miles (22 km) south of Santa Fe that's known for its scenic dry creeks, sandstone formations and vast savanna grasslands, and set aside for hiking and riding.

Above, Katy saddles up her horse, Kleo, to ride through an intermittent stream bed—locally known as an arroyo.

"We are the single biggest users of the ecosystem . . . We have a responsibility."

Chaps protect the family's legs against
the harsh branches and thorns that are
common in the arid landscape, while cow-
boy hats keep the sun off their heads and
the rain from seeping down their backs.

RON FINLEY

By sowing seeds in the barren public spaces of South Central Los Angeles, Ron Finley reaped an unexpected reward: a newfound calling as a community activist. With each project that takes root, his motivation only grows.

Many who garden find their work restorative. Rarer are gardeners whose efforts have sparked political awakenings. Ron Finley grew up in the blighted South Central region of Los Angeles and went on to become a successful fashion designer and personal trainer. But it was an act of gardening that crystallized his political consciousness.

In his 2012 ten-minute TED Talk (which has now surpassed 3.5 million views), Finley recounts the story of his political awakening in a series of pithy, and often delightfully unprintable, turns of phrase. South Central, he says, is what's known as a food desert—a term often used to describe inner-city areas where the only food options are fast-food chains and dollar stores. Disheartened by his community's limited access to fresh fruits and vegetables and the resulting sky-high rates of obesity, hypertension and other diet-related health problems, Finley transformed his parkway—West Coast terminology for the planted portion of a sidewalk—into a garden with vegetables and banana trees.

As a result, the city of Los Angeles issued a citation, then a warrant for Finley's arrest, on the grounds that he was working public land without a permit. The public outcry that ensued successfully changed the law in Los Angeles that had prevented people from gardening on parkways. It also propelled Finley into a pioneering new career: community gardening activist.

"Gardening is the most therapeutic and defiant act you can do, especially in the inner city," Finley says on the phone

from his new permanent project space in Los Angeles. "I've witnessed my garden become a tool for education, a tool for the transformation of my neighborhood."

Finley is adamant that home gardening has the opportunity to transform more than just his block in South LA. An increase in individuals' self-sufficiency can also positively disrupt the social and political systems that perpetuate self-defeating cycles in low-income communities.

"Just think about even one percent of us starting to grow our own food," Finley says. "Think how much money that would take out of the system, from healthcare to grocery stores. People growing their own food is dangerous [to the status quo]." Finley's number one tip to novice gardeners is indicative of his straight-talk approach: "Plant what you like to eat. Don't plant no shit you don't like."

But growing food to eat isn't Finley's only motivation to keep on mulching. "I'm not always planting for production. I also plant for beauty, for engagement. My garden is basically a big-ass social experiment. I'm an urban sociologist asking the question, 'How do people engage with something they're not used to seeing in the urban environment?'"

Finley's 'Russian Mammoth' sunflowers, in particular, have caused quite the stir in the neighborhood. The plants can stand ten feet (3 m) tall, supporting flowers over a foot (30 cm) in diameter. "Kids stop and ask, 'Yo, is that real?' People have never seen anything like this. It's that kind of engagement I want."

Finley's edible community garden seeks
to tackle inner-city food scarcity, which
he experienced himself (he recalls having
to drive forty-five minutes to buy a fresh
tomato). His project has been fruitful: His
garden now yields Cape gooseberry
(*Physalis peruviana*), pictured above, and
kumquats, pictured opposite.

"I've witnessed my garden become a tool for education,
a tool for the transformation of my neighborhood."

The project is built around a drained, graffitied swimming pool at the edge of a train track. Pictured below are containers of variegated agave and aloe and a fig tree (*Ficus carica*). Finley also keeps a beehive on-site.

Finley's plants include devil's bell flower (*Datura stramonium*), burro and ice cream banana trees, hen and chick succulents (*Sempervivum*), agave, mother-in-law's tongue (*Dracaena trifasciata*), nectarine and pomegranate trees, and jade plants (*Crassula ovata*).

ANJA

CHARBONNEAU

Anja Charbonneau has built a sisterhood around cannabis. As the founder of *Broccoli* magazine, she is fostering intelligent conversation around the plant's use and disseminating a sharper, more stylish understanding of "high culture."

The legalization of cannabis in many parts of the world is lifting a cloud of secrecy around its use. The resulting cultural shift toward a more open, comfortable conversation about the plant has found physical form in *Broccoli*, a magazine "created by women who love weed."

It was launched in 2017 in Oregon—three years after the state lifted the ban on recreational use of the drug—by "longtime weed lover" (and former creative director of *Kinfolk*) Anja Charbonneau. *Broccoli* sets itself apart from the publications that previously dominated this niche: "Magazines like *High Times* have been around forever, but they present a very masculine version of the culture," Charbonneau says. "Knowing all these interesting, creative women who liked weed, I thought we needed to have something, and I believed there would be others out there who needed to see it."

The gamble paid off: As well as attracting a host of advertisers keen to reach the new market of legal cannabis users, the magazine has quickly reached an international audience delighted to see their positive relationship with the plant reflected in its pages. "It ended up touching a lot of different people that I wasn't expecting," Charbonneau says. Building on its success, the magazine has expanded its remit to offer events, workshops and community meetups to bring people together, as well as a podcast to continue the conversation between print issues and reach consumers the magazine does not.

Aesthetically, the magazine performs a neat balancing act between the trippy visuals long associated with cannabis and the refined graphics expected from a design-conscious publication; it taps into "high culture" in both senses of the word. One example is the cover story of its first issue: a shoot with cannabis plants in the style of Japanese ikebana flower arrangements. "The pot leaf has a cheesy reputation associated with stoners in their basements or bad T-shirts at the beach, but people have forgotten that it's a very elegant and beautiful plant," Charbonneau explains.

In *Broccoli*, cannabis is less an editorial focus than a jumping-off point for telling interesting tales. One issue, for example, includes an interview with the late music producer and songwriter Allee Willis, who co-wrote the theme song for the sitcom *Friends* and was known for throwing decadent parties. Another feature goes inside a summer camp for children from Latin American countries that have felt the impact of the war on drugs.

The latter is a reminder that cannabis is, of course, still illegal in most places. To that end, the central mission of the publication seems to be to fly the flag for cannabis and its community as a positive force. "Cannabis is a strong connector," Charbonneau says. "I've heard of so many people trying it for the first time because something traumatic happened in their life, and interesting stories of transformation. When you experience something profound like that, you can bond with others over those shared experiences. It's a unique thing."

For the launch issue of *Broccoli*, floral designer Amy Merrick created hemp ike-bana arrangements, which Charbonneau photographed. The editor says she is often to be found reading in this spot: "I like to find interesting old photo books at used book stores," she says.

Charbonneau lives in Portland, near the
Lan Su Chinese Garden. A collaborative
public project between the city and Suzhou
in China's Jiangsu province, the garden
is best known for its plum blossom trees
(*Prunus mume*).

The cover of *Broccoli's* Spring 2020 issue showed a flower smoking a joint, shot and styled by Carl Ostberg. "We often show flowers in our work. It helps to make the idea of cannabis accessible and comfortable for people, because everyone can understand flowers," says Charbonneau.

SKOGSKYRKOGÅRDEN

Beautifully designed graveyards often double as communal gardens for the living. On the outskirts of Stockholm, the wild beauty of *Skogskyrkogården* cemetery has been providing sanctuary to the local community for the last hundred years.

Beneath the reaching branches of twelve old elms, calm spreads over the walkways of Skogskyrkogården—a Unesco World Heritage cemetery just south of Stockholm. The faint perfume of the grounds' flowers mingles with the wilder scent of its pine trees. Rough granite walls set within a hilltop grove enfold a shared space of grief and memory for the local community. Bright against the forest below, smooth sandstone walls shape the Chapels of Faith, Hope and the Holy Cross; three weathered smokestacks rise together behind the chapels.

When Skogskyrkogården was founded at the beginning of the last century, cemeteries were generally considered to be "gardens of the dead"—grandiose mementos to the deceased. Architects Gunnar Asplund and Sigurd Lewerentz proposed that the site shun spectacle and instead complement the existing landscape, blending nature and architecture into a seamless whole.

Asplund and Lewerentz won the initial competition for Skogskyrkogården cemetery in 1915 and continued to develop its buildings and grounds over the following three decades. Their architectural work carried echoes of an ancient past into the twentieth century, blending classical elements with a solemn sort of Nordic modernism. Their buildings, and the vast expanse of lawns and forest around

them, created a "spiritual landscape," architectural historian Caroline Constant explains. Skogskyrkogården, she says, was designed as a place that "builds upon the common private experience of mourning to convey a sense of oneness with humankind and with nature."

Today, mourners still walk together toward the chapels on a flagstone path that climbs from the cemetery entrance past an enigmatic Tuscan portico overhung with Swedish whitebeam trees. It frames glistening rock stained a primordial red through a slow process of oxidation. The path rises up a sloping lawn toward a large granite cross set against the sky, from which the slow procession continues to a low ridge where the view opens southward toward a grove of birches planted in rows before a high mass of conifers. Many gravel paths fork out from here into the tall pines. Along the open forest floor, a hundred thousand low gravestones extend in solemn rows. Small beds of annual flowers at each brighten the gray regularity. The gray columns of the Chapel of the Resurrection's Corinthian porch echo the silhouettes of surrounding tree trunks.

The meditation grove, shaded processional routes, forest burial grounds, and chapel glades seem to bind primal forest and cultivated land together in a place where spirits of departed companions can abide peaceably with the living.

Built on and around the rolling contours of a former gravel pit, the graveyard is intentionally unregimented. Graves are reserved for the woodland without excessive alignment, inspired by ancient and medieval Nordic burial traditions. The granite cross, pictured left, was added by Asplund in 1939.

The Woodland Crematorium, pictured opposite, was designed according to the functionalist principles popular in 1930s Sweden. Asplund placed waiting rooms and gardens between its three chapels, allowing visitors space to mourn with views of the landscape.

Five Tips

Gardens have long brought people together. As the seeds you sow or the plants you tend grow and thrive, so too will opportunities to share the fruits of your labor. The following pages offer some advice on how to create a community—or else contribute to one—with some simple gardening skills. Whether sharing flowers and cuttings from your garden or finding ways to get involved in growing when you don't have access to your own patch of green, these tips will help to bolster old bonds and let new ones blossom.

How to Plant Some Roots

Words by Melissa Mabbitt
Gardening writer, editor and hands-on horticulturist

The code entwined in a bouquet was widely understood just a few decades ago, but floriography, as it is known, is a language that has faded from its Victorian height. Only a few remembrances linger—red roses mean love, an olive branch offers peace. Birthday cards, telephone calls and texts are quick and casual, but there is room, surely, for a more nuanced and imaginative way to express a heartfelt sentiment through a message picked out in blooms.

Daffodils are a sign of new beginnings, and a bunch of their cheerful golden faces will light up a new home or send hope and joy to a new parent. Send them to a friend starting a new job, too, along with primroses, whose bright petals and sturdy, unbending stems represent confidence.

For a friend who is not so lucky—losing out on a job or promotion—irises of any sort encourage bravery. A modern twist would be to send a cactus: What other plant can so thoroughly signify resilience?

In the face of illness, a posy of fragrant herbs is said to aid recovery, especially sage and chamomile. Feverfew offers warmth and fennel strength, while brighter flowering *Salvia* is a pick-me-up for loved ones on the mend.

Whether a cryptic communication or something as obvious as a deep red rose for passionate love, pick out your message in the code of flowers and it will be understood intuitively.

2. HOW TO TALK TO PLANTS

Most gardeners love to garden because they find solace in their plants. These living entities ask nothing of us other than to be cared for and, by caring for them, plants become like friends or children. Perhaps that's why it's common to see gardeners refer to their houseplants as their "babies" online.

It's not much of a leap to imagine this intimacy expressed in words and song. For many years, the idea of talking to plants has been scoffed at, and yet many gardeners like to do it, perhaps (quietly) whispering encouragement or threatening the compost heap if flowers underperform. Darwin himself investigated this idea, and though many studies have been carried out with inconclusive results, one study by the UK's Royal Horticultural Society seems to suggest plants respond well to female voices. Evidence also suggests that plants do indeed "talk" to each other, sending signals under the soil.

It has also been suggested that the carbon dioxide we breathe onto leaves encourages plant growth, but in reality, talking or singing to your plants could be good for them for one simple reason: It means you will spend more time with them, notice problems and so be more likely to look after them well. So sing, recite, nag or whisper—just do it while you are lovingly tending to your plants' needs.

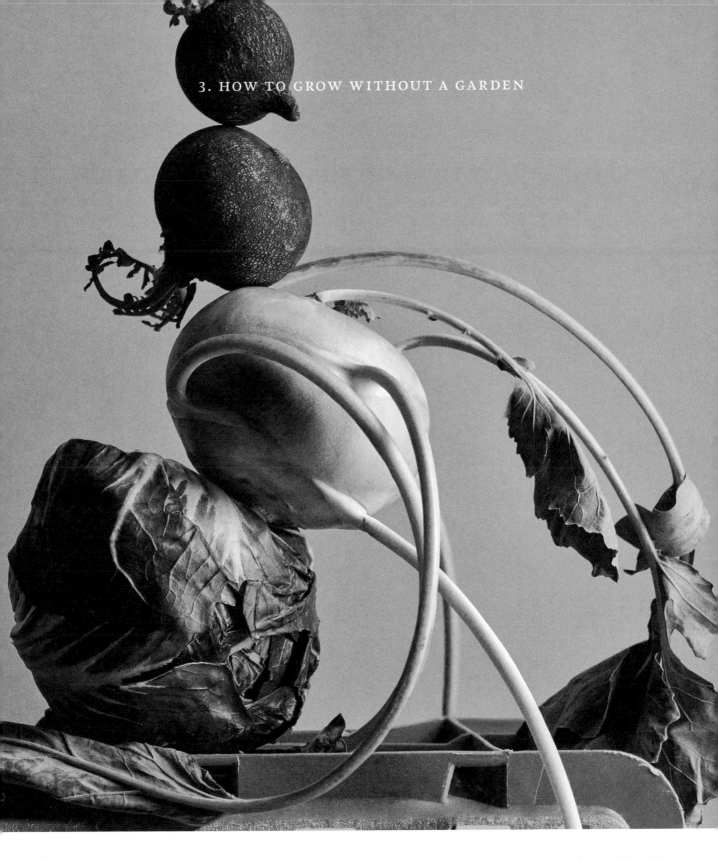

Even if you don't have a sprawling yard in which to plant a vegetable garden, that doesn't mean fresh homegrown food is off the menu. Salad vegetables such as lettuce, radish and baby beetroots are the easiest to grow in small containers; a window box, wall planter or any pot positioned where plants will receive good light can be used as a growing space. Make sure your containers are lightweight and have plenty of drainage holes.

Smaller cultivars of fruit and vegetables can be grown even in tiny spaces. Try hanging basket tomatoes such as 'Tumbling Tom Red' or 'Tumbling Bella.' They'll need daily watering as hanging baskets can dry out quickly, plus a weekly feed with a liquid tomato fertilizer. Check to

make sure your tomatoes get at least six hours of sun each day during summer to help the fruits ripen. Smaller cane fruits such as 'Ruby Beauty' raspberries and 'Opal' blackberries are also ideal for container cultivation. Strawberries of any sort always make an ideal candidate for hanging baskets or wall planters.

Many edible plants make ideal houseplants. Sweet potatoes put out trailing vines with beautiful heart-shaped leaves that can be trained artfully over bookshelves or up a wall. Small citrus trees and pineapple plants will give you bite-sized delicacies indoors, and a tender hibiscus will provide beautiful and edible flowers with which to decorate cakes and cocktails.

4. HOW TO SHARE GARDEN CUTTINGS

The best time to take a cutting is when nobody's looking, as the old joke goes. Jesting aside, gaining a cutting is more likely to be an act of friendship than thievery. A plant grown from a cutting is one that will always remind you of the place or person who gave it to you.

The most common cuttings are "softwood" cuttings, which are newly sprouted young shoots that are green, leafy and flexible (hence "soft"). Take them in late spring or early summer, and use clean secateurs to clip a few shoots between two to four inches (5 to 10 cm) long from your chosen plant. They should be leafy, without any flower buds. Trim off all but the top two leaves, taking care not to tear or nick the stem, and then trim the cut end to just below a leaf node (a notch where a leaf sprouted from the stem). If you receive a cutting, seal it in a plastic bag with a spritz of water to keep it damp while you transport it home.

Put the cutting in a container of water. When it begins to develop roots, fill a pot with compost mixed with a handful of gravel. Push the cutting into the pot, water lightly, and put a plastic bag over the pot, securing it with an elastic band. Place the pot on a bright windowsill, checking every few days to make sure the compost doesn't dry out. In a few weeks, the roots will become established and the cutting will have transformed into a baby plant.

5. HOW TO GUERILLA GARDEN

Guerilla gardening hit the headlines nearly fifteen years ago when city-dweller and frustrated gardener Richard Reynolds started late-night expeditions to green up neglected London planters, verges and roundabouts.

You can join a local group or go responsibly rogue. Start close to home. Are there neglected concrete planters near your building's entrance? Do some of the street trees have room for a few lavender plants at their feet? The closer the project is to your home, the easier it will be to do the necessary maintenance as the plants become established.

Planting pockets often form in old walls, and there may be cracks where you can squeeze in a few goldmoss stonecrop (*Sedum acre*) or *Sempervivum*, which will decorate the brickwork with rich color and require next to no upkeep. Perhaps the easiest way to guerilla garden is to scatter seeds in the spring, wherever there is a patch of bare ground. Rake the soil lightly and sprinkle on fast-growing and drought-tolerant annuals such as any poppy, nasturtiums, yarrow or pot marigolds.

Bear in mind that gardening any plot of land you do not own may be illegal, but in some city areas, such as Hammersmith and Fulham in London and in Munich, councils welcome improvements to common land and even offer advice on how to do it. It pays to check first with a quick browse online or a phone call to your local authority.

Books

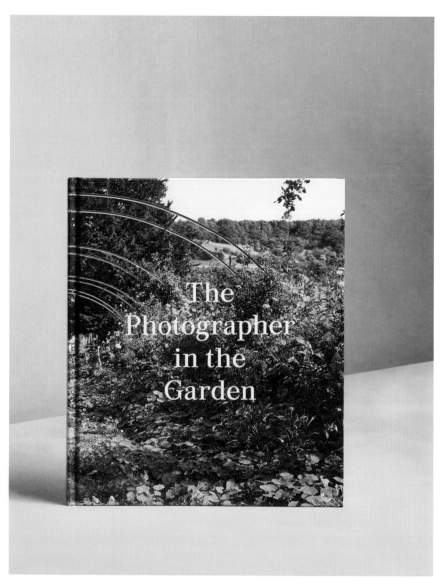

Gardens—and the pride they instill in us—can be shared with others through photographs. *The Photographer in the Garden* traces the garden's rich history in photography, offering inspiration for how to capture its glory.

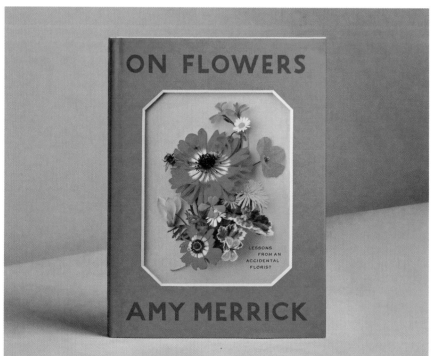

Bouquets are ephemeral and often expensive gifts, whereas advice on how to arrange flowers is a gift that keeps on giving. In *On Flowers: Lessons from an Accidental Florist*, Amy Merrick shares her hard-won knowledge on the art.

Gardens can be a direct extension of the kitchen, providing ingredients and inspiration for countless dishes. *The Garden Chef* includes more than 100 imaginative garden-focused recipes. Ideally, serve them alfresco.

Barcelona-based *The Plant* magazine takes an in-depth look at plants through the lens of the creative communities who love them.

Gardenista is an all-in-one manual featuring hundreds of tips on how to make your outdoor space as welcoming to guests as your living room.

THANK

YOU

The *Kinfolk* team would like to wholeheartedly thank each of the people featured in these pages for taking the time to welcome us into their lives, gardens and studios. We're grateful for your patience throughout the process and for being so accommodating with your time and generous with your hard-won knowledge.

We would also like to sincerely thank all of the talented photographers and writers around the world who brought those stories to life so beautifully. As always, we consider it an honor to collaborate and publish your work. Particular thanks to Darryl Cheng, Amy Merrick and Melissa Mabbitt for offering their tips, advice and expertise, and to Lauren Boudreau and Sandie Lykke Nolsøe for imagining and capturing them so beautifully.

The Kinfolk Garden's creative team is made up of John Burns, Staffan Sundström, Harriet Fitch Little, Julie Freund-Poulsen and Gabriele Dellisanti. In-house production has been the work of Susanne Buch Petersen and Eddie Mannering—we'd like to extend particular thanks to them for keeping us all on track and on time.

Thanks also to the rest of our colleagues at *Kinfolk*: Christian Møller Andersen, Cecilie Jegsen, Alex Hunting, Chul-Joon Park, Seongtaek Jang and Nathan Williams. A thank-you to Amy Woodroffe for her feedback.

Thanks to model Sarah Runge—the woman on the cover of this book—and to cover photographer Sarah Blais, stylist Anne Törnroos, hair and makeup artist Sabine Simmelhag and set designers Johanne Aurebekk and Cassandra Bradfield for producing the beautiful cover image.

We would like to thank our publisher, Lia Ronnen at Artisan Books, for her continued support of *Kinfolk* and for her ideas, advice and feedback throughout this project. We'd also like to extend sincere thanks to Artisan team members Bridget Monroe Itkin, Zach Greenwald, Nancy Murray, Bella Lemos and Suet Chong for their instrumental feedback and efforts, and to Theresa Collier, Allison McGeehon and Amy Kattan Michelson for bringing the book to life.

Last, but by no means least, we would like to thank our readers for your continued support. We hope this book has inspired you to introduce more nature into your daily life.

Credits

COVER

PHOTOGRAPHY
Sarah Blais

STYLIST
Anne Törnroos

SET DESIGN
Johanne Aurebekk &
Cassandra Bradfield

HAIR & MAKEUP
Sabine Simmelhag

MODEL
Sarah Runge

———

TIPS

PHOTOGRAPHY
Staffan Sundström

PHOTOGRAPHY — PRODUCTS
Sandie Lykke Nolsøe

ARRANGEMENTS & SET DESIGN
Lauren Boudreau

WORDS — CARE
Darryl Cheng

WORDS — CREATIVITY
Amy Merrick

WORDS — COMMUNITY
Melissa Mabbitt

PHOTOGRAPHY

56 – 65, 76 – 81
Jonas Bjerre-Poulsen

COVER, 272 – 283
Sarah Blais

11, 28 – 41, 110 – 117, 166 – 175,
262 – 271, 284 – 293, 294 – 303, 346 – 348, 350
Rodrigo Carmuega

6, 92 – 101, 156 – 165, 316 – 323
Justin Chung

194 – 201
Claire Cottrell

146 – 155
Cecilie Jegsen

66 – 75
Renée Kemps

12, 16 – 27
Ekin Özbiçer

102 – 109
Cecilia Renard

8 – 9, 138 – 145, 216 – 227, 332 – 337
Staffan Sundström

128 – 137, 324 – 331
Kourtney Kyung Smith

202 – 215
Dominik Tarabanski

254 – 261
Benjamin Tarp

42 – 55, 82 – 91
Zoltán Tombor

182 – 193
Gustavo García-Villa

5, 240 – 253
Alexander Wolfe

304 – 315
Corey Woosley

176 – 181
Yuna Yagi

WORDS

156, 176, 332
Alex Anderson

138, 146
Rima Sabina Aouf

13, 349
John Burns

216
Gabriele Dellisanti

42, 82
Daphnée Denis

66
Timothy Hornyak

102, 182, 294
Scarlet Lindeman

56, 76, 324
Debika Ray

202, 304
Ben Shattuck

92, 128, 240, 316
Stephanie d'Arc Taylor

16, 28, 194, 254, 262, 272, 284
Pip Usher

110, 166
Annick Weber

Library of Congress Cataloging-in-Publication Data

Names: Burns, John, 1968-   author.
Title: The kinfolk garden : how to live with nature / John
    Burns.
Description: New York, NY : Artisan Books, a division of
    Workman Publishing Co., Inc., 2020.
Identifiers: LCCN 2020020282 | ISBN 9781579659844
    (hardcover)
Subjects: LCSH: Gardens—Design. | Landscape design. | Flower
    arrangement. | Indoor gardens. | Gardens—Pictorial works.
Classification: LCC SB465 .B92 2020 | DDC 712/.6—dc23
LC record available at https://lccn.loc.gov/2020020282

PUBLISHED BY
Artisan
A division of Workman Publishing Co., Inc.
225 Varick Street
New York, NY 10014-4381
artisanbooks.com

Artisan is a registered trademark of
Workman Publishing Co., Inc.

Published simultaneously in Canada by
Thomas Allen & Son, Limited

Printed in China

First printing, October 2020

10 9 8 7 6 5 4 3 2 1